The ArcGIS® Book

10 Big Ideas about Applying The Science of Where™

Esri Press
REDLANDS | CALIFORNIA

Esri Press, 380 New York Street, Redlands, California 92373-8100

Printed in the United States of America
21 20 19 18 17   1 2 3 4 5 6 7 8 9 10

Ask for Esri Press titles at your local bookstore or order by calling 800-447-9778, or shop online at esri.com/esripress. Outside the United States, contact your local Esri distributor or shop online at eurospanbookstore.com/esri.

Esri Press titles are distributed to the trade by the following:

In North America:
Ingram Publisher Services
Toll-free telephone: 800-648-3104 - Toll-free fax: 800-838-1149
E-mail: customerservice@ingrampublisherservices.com

In the United Kingdom, Europe, Middle East and Africa, Asia, and Australia:
Eurospan Group
3 Henrietta Street, London WC2E 8LU, United Kingdom
Telephone: 44(0) 1767 604972 - Fax: 44(0) 1767 601640
E-mail: eurospan@turpin-distribution.com

All images courtesy of Esri except as noted.

On the cover:
This map shows the aggregated locations of ArcGIS Online service requests over a moving 24-hour window, on a software release day in September 2016. It paints a global picture of the vast and unifying community of geospatial technology users. Each hexagonal grouping is further split into three diamond-shaped sections to provide a more geographically nuanced look at a clustered phenomenon. The basemap is a hillshade effect generated from the TopoBathy elevation web service. The map was processed and assembled entirely in ArcGIS Pro.

Contents

Using data derived from the NASA Earth at Night imagery set, this map looks at where night lights have turned on (blue) or dropped out (magenta) over the past 4 years in Africa, Europe, and Asia.

Introduction: Applying The Science of Where

The Web GIS revolution is radically altering how information about the world around us is applied and shared.

This is a book about ArcGIS, the Web GIS platform. But ArcGIS is more than just mapping software that's running online. It's actually a complete system for discovering, consuming, creating, and sharing geographic data, maps, and apps designed to fulfill particular objectives.

The twin goals of this book are to open your eyes to what is now possible with Web GIS, and then spur you into action by putting the technology and deep data resources in your hands via the QuickStarts and Learn ArcGIS lessons that are included in each chapter. By the end, if you complete all the exercises, you'll be able to say you published web maps, used story maps, built a 3D cityscape, configured a custom web app, performed sophisticated spatial analysis, and much more.

The basics of ArcGIS are easy, engaging, and fun, and even more sophisticated features (such as spatial analysis and web app development) are now accessible to everyone, not just the experts. With the world's geography at your fingertips, you'll be empowered to effect positive change in the world around you.

Once the exclusive realm of technologists, digital mapmaking has gone mainstream, empowering everyone.

Freely available and approachable, Web GIS makes for a kind of democratization of mapping and analysis of the world around us. If we think of geography as the ultimate organizing principle for the planet, then Web GIS is the operating system. The challenges we face, from our local neighborhoods to our world as a whole, all share the common tenets of geography: they are happening somewhere, which places them squarely "on the map."

The Science of Where:
Unlock Data's Full Potential

How this book works

See this book come alive at www.TheArcGISBook.com

You are reading the print version of this book. It exists wholly online at TheArcGISBook.com. Bookmark it now for when you're ready to sit down at your computer and work with ArcGIS.

What's new in the second edition?
This new edition includes over 250 new examples, all new Learn ArcGIS lessons, and an exclusive membership in the Learn Organization in ArcGIS Online.

Who is the audience?
This book has been designed with several audiences in mind. The first is the professional mapping community—the people who create or work with geospatial data as a dedicated activity—in particular, those GIS (geographic information systems) professionals who are just beginning to leverage mapping online. The second is the broader world of web technologists, information workers, web designers, and Internet-savvy professionals in many other fields. The technology has become so ubiquitous and easy to use that a third audience is any individual with an interest in maps and ideas for how to apply them. The only prerequisites are a desire to better understand web GIS and a roll-up-your-sleeves attitude.

Learn by doing
This is a book that you *do* as well as *read*, and all you really need is a personal computer with web access. The adventure starts when you engage yourself in the process by completing the lessons in this book. Each step of the way, you will gain new skills that take you further. Mapping professionals are in high demand for a reason. Businesses, governments, and organizations of all stripes can see the value. This book is a call to action and a blueprint for how to get there. It's about applying geography to your specific situation, problem, or challenge, and finding a solution with GIS.

While reading this book on one of several available platforms, including print, you can practice making maps on the web with your computer. With the web version, you will experience and use many of the example maps and apps as they come to life on the screen.

In each chapter, the QuickStarts tell you what you need to know about the software, data, and web resources that pertain to that aspect of the ArcGIS system. The Learn ArcGIS lesson pages are your gateways to online instructional content.

Although structured with one big idea per chapter, each chapter provides many more granular ideas. Open the book and read any page, or read it front-to-back and be part of the adventure every step of the way. Experience web GIS at your own pace according to your own interests.

Most importantly, we want you to feel empowered to dive right into ArcGIS technology and expand your horizons by doing real mapping and analysis using web GIS. What problem in your life or within your purview would you like to investigate? If it has a geographic element (and most do), then it's something you can tackle with GIS.

Learn ArcGIS

Designed for self-study with lessons covering many GIS applications, Learn ArcGIS presents a new way for you to understand GIS concepts and technology: putting the story and problem first. With these lessons covering the entire breadth of the ArcGIS platform, you'll learn how to use a variety of different applications and techniques to help solve real-world geographic problems.

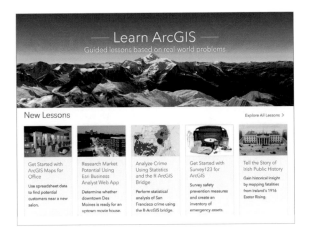

ArcGIS membership and ArcGIS Pro license

Most lessons require membership in an ArcGIS Online organization. If you are already a named user with publisher privileges in an ArcGIS organization, you can complete all the ArcGIS Online lessons. Data links are provided in each lesson. If you're not yet a member of any ArcGIS organization, you can join one called Learn ArcGIS, which is designed for students and self-study learners. Once you're signed into this special learning organization, you can use the link on page 15 to download and install ArcGIS Pro on your local computer. With this limited-time membership, you get an ArcGIS Online account and an ArcGIS Pro license. Your account and content will be deleted after 60 days.

Instructional Guide for The ArcGIS Book

Written by GIS instructors Kathryn Keranen and Lyn Malone, the *Instructional Guide for The ArcGIS Book,* second edition, serves as a companion to *The ArcGIS Book,* second edition, and includes relatable activities and lessons that correspond to each chapter of *The ArcGIS Book.*

GIS students and seasoned pros will hone their GIS skills while they build and publish web map apps; use live data feeds in apps; communicate information using maps; create and share Esri Story Maps, answer complex questions using web maps and analysis tools; and make 3D map presentations. Some lessons require no software at all, while others call for using ArcGIS Online or Esri's cloud-based GIS apps.

Whether you are a self-learner, currently teaching, or are planning to teach GIS, the *Instructional Guide for The ArcGIS Book* provides the materials to explore and apply GIS concepts and ArcGIS tools.

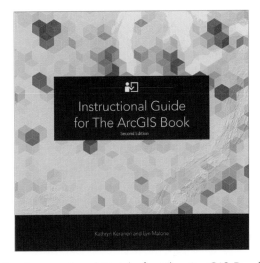

The Instructional Guide for The ArcGIS Book, Second edition is available at no cost in PDF format at TheArcGISBook.com.

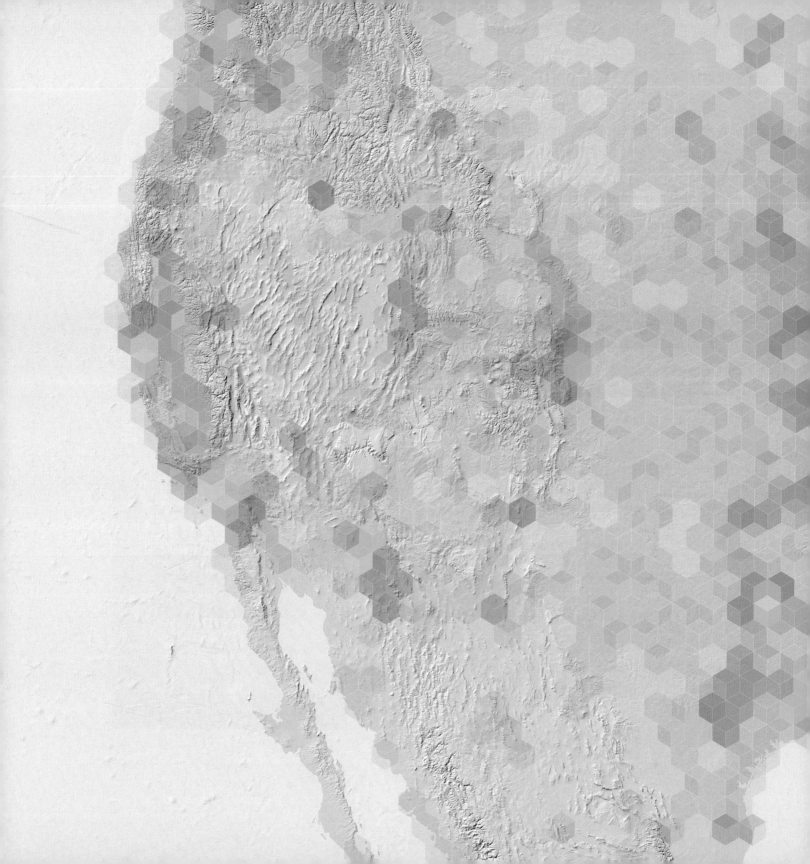

GIS Provides a Common Visual Language

Transforming our understanding of the world

Maps and data underpin GIS, a technology that organizes information into all types of layers that can be visualized, analyzed, and combined to help us understand almost everything about our world. Web GIS connects and organizes many individual GIS systems into a collective GIS for the planet—available for use by anyone with an Internet connection. It's time to join and participate.

It's a Web GIS world

Mapping and analytics connect everyone

GIS has the extraordinary potential to touch every web-connected being through a common visual language that unites people across organizations and throughout the world. Today, hundreds of thousands of organizations in virtually every field of human endeavor are using GIS to make maps that communicate, perform analysis, share information, and solve complex problems. This is literally changing the way the world works. The power of this shift in the ways that people think about maps and geographic data is evident in this small gallery, which highlights some compelling examples.

GIS maps are uniquely of the moment

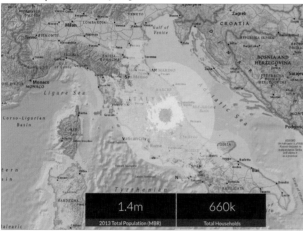

Published moments ofter the event, this shakemap portrayed the extent of area and people affected by the devastating 2016 earthquake in central Italy.

GIS maps reveal rapid global change

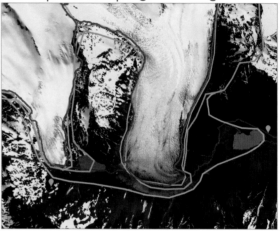

If a picture tells a thousand words, a map tells a thousand pictures. This map starkly reveals the disheartening extent of glacier retreat in the Southern Hemisphere.

GIS maps illuminate natural phenomena

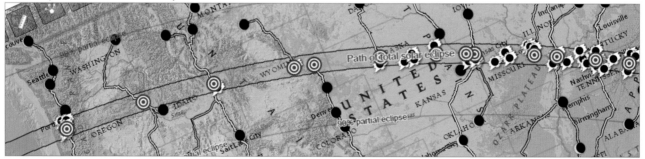

This interactive web app shows the path of the August 21, 2017, total eclipse of the sun. With totality literally reaching from coast to coast in the United States, this is the kind of information that is ideally suited for a map-based display.

GIS maps illuminate social issues

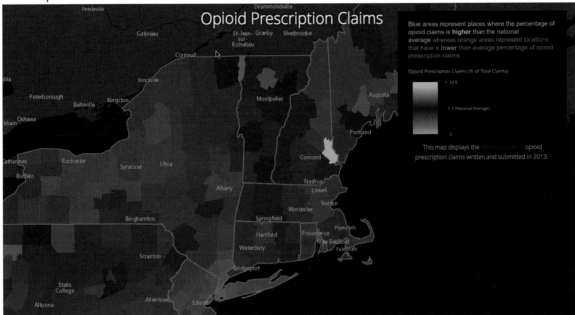

Social issues are driven by geography at global, national, and local levels. This interactive map of opioid prescription claims (zoomed into the Eastern Seaboard of the United States), reveals the subtle and often tragic geographic patterns that emerge when data is well-mapped.

GIS maps enable smart planning

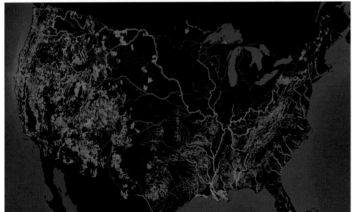

Green infrastructure is a method for addressing urban and climatic challenges by building with nature. It is a way to view the land around us in a more holistic ecological manner.

GIS maps manage large events

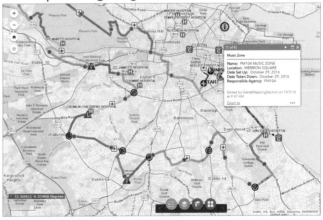

This useful app details the key particulars of each segment on the route of the Dublin Marathon. The public now expects to find this depth of information for events of all kinds.

Thought leader: Jack Dangermond
GIS: Understanding The Science of Where

GIS is about uncovering meaning and insights from within data. It is rapidly evolving and providing a whole new framework and process for understanding. With its simplification and deployment on the web and in cloud computing as well as the integration with real-time information (the Internet of Things), GIS promises to become a platform relevant to almost every form of human endeavor—a nervous system for the planet. This system is now not only possible, but in many ways we believe it's inevitable. Why?

GIS integrates data about everything—and, at the same time, it provides a platform for intuitively understanding this data as an integrated whole. This GIS nervous system is providing a framework for advancing scientific understanding, and integrating and analyzing all types of spatial knowledge (all the "-ologies" such as biology, sociology, geology, climatology, and so on).

GIS provides a platform for understanding what's going on at all scales—locally, regionally, and globally. It presents a way to comprehend the complexity of our world as well as to address and communicate the issues we face using the common language of mapping. At Esri, we refer to this idea as The Science of Where.

Our world is increasingly being challenged by expanding populations, loss of nature, environmental pollution, and the increasing dilemma of climate change and sustainability. My sense is that humans have never been more capable of sharing and addressing these issues. My belief is that GIS not only helps us increase our understanding, but it also provides a platform for

Jack Dangermond is president and founder of Esri, the world leader in GIS software development and its application in business, health, education, conservation, utilities, military and defense, oceanography, hydrology, and many other fields.

 Video: GIS-Enabling a Smarter World

collective problem-solving, decision-making, and perhaps most critical of all, collaboration. However, it's going to take all our best people, our most effective methodologies and technologies— scientists from many disciplines, our best thinkers, all our best design talent collaborating to create a sustainable future. GIS technology and GIS professionals will play an increasingly important role in how we respond to and confront our collective problems.

My hope is that by using GIS to apply The Science of Where, we can discover deeper insights, make better decisions, and take them to action.

Enabling a smarter world
GIS provides a framework and process

The Internet of Things is becoming real. We're learning how to measure virtually everything that moves and changes on the planet with a web of connected instruments that are gauging water flows, documenting changes in climates, and pinpointing where people and things exist. And this lattice of information is becoming available through the Internet, basically instrumenting our planet with GIS. Location mapping and GIS is becoming an essential framework or language to help us keep track of what's going on.

When fully realized, this framework allows a sort of virtuous cycle. GIS users, working in their localities—each contributing their small piece of the larger geographic puzzle—repeatedly apply the science of our world to collect and organize their geographic information into effective representations. They use maps along with interactive charts and graphs to visualize, portray, and share their results. With GIS, users go even further and ask deeper, more probing questions to model and test different scenarios. These GIS practitioners apply The Science of Where to inform the way that people make decisions and take action. Think about GIS users repeating this cycle over and over in hundreds of situations and locations worldwide—it adds up to science in action.

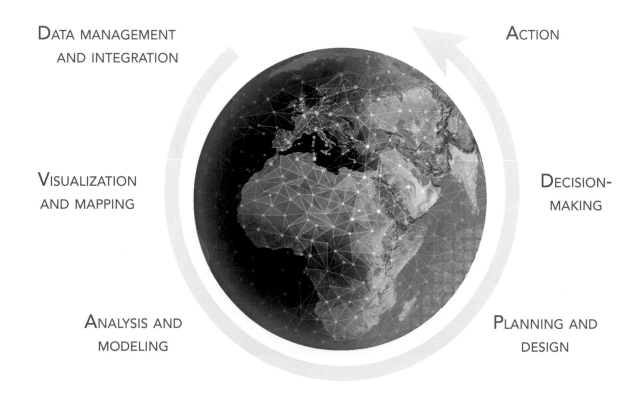

DATA MANAGEMENT AND INTEGRATION

ACTION

VISUALIZATION AND MAPPING

DECISION-MAKING

ANALYSIS AND MODELING

PLANNING AND DESIGN

It all begins with a map

Web GIS revolves around the map. It's the framework for your data and the primary geographic container that gets shared and embedded in your apps. In ArcGIS it is called a "web map." The purpose of the web map below is simple enough: to show the last 120 days of earthquakes everywhere on Earth. (By the way, if you're reading the print edition, make sure you access the book on your computer as well to get the full, up-to-date experience.)

There are several points of interest right here on this map. First of all, it's navigable, which means you can pan and zoom. The map has many zoom levels built in, each one revealing more detail the closer you get. Click on any earthquake symbol to learn the magnitude and date of each event.

These little windows of information are called "pop-ups," and by the time you finish this chapter, you'll know how to configure them.

The map also has scaled symbols, showing the relative magnitude of each earthquake. The background map is symbolized as well, in this case in muted dark tones that set off the bright earthquake symbols.

This data, organized with this combination of symbology, reveals an interesting pattern: the well-known Ring of Fire. This map could be easily embedded on any web page or in an app. But where did it originate? It began life as a web map in the ArcGIS map viewer.

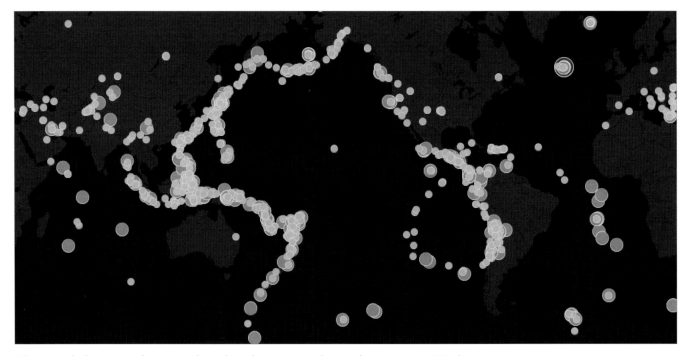

This simple live map shows earthquakes that occurred over the previous 120 days.

The expansive reach of Web GIS

Across organizations and beyond

The geographic organizing aspect of GIS has been part of the thinking from the beginning, but now factor in the impact of the web. Web GIS provides an online infrastructure for making maps and geographic information available throughout an organization, across a community, and openly on the web. This new vision of Web GIS fully complements, integrates, and extends the work of existing GIS professionals.

Web access to data layers is straightforward: every layer has a web address (a URL) making it easy to locate and share online. Because every layer is georeferenced, Web GIS becomes a system for integration that facilitates the access and recombination of layers from multiple providers into your own apps. This is significant for the millions of GIS professionals worldwide who are building layers that serve their individual purposes. By simply sharing these layers back into the online GIS ecosystem, they are adding to a comprehensive and growing GIS for the world. Each day, this resource grows richer and is tapped by ArcGIS users and shared on the web.

GIS is continually evolving. Its information model was originally centered around local files on a single computer. From there, it evolved into a central database environment based on clients and servers. The most recent evolutionary stride has taken it to a system of distributed web services that are accessible in the cloud. ArcGIS is now a Web GIS platform that you can use to deliver your authoritative maps, apps, geographic information layers, and analytics to wider audiences. You do this by using lightweight browser clients and custom applications on the web and on smart devices, as well as desktops, as you'll see in later chapters.

These three layers depicting Harvard University facilities, global airport locations, and Dutch historical maps, respectively, are among thousands available on ArcGIS Online.

Much of the work of traditional GIS users and experts has involved building and maintaining key foundational layers and basemaps—information products that support a particular mission. Huge investments have been made to compile these basemaps and data layers in great detail and at many scales. These include utility networks, parcel ownership, land use, satellite imagery and aerial photography, soils, terrain, administrative and census areas, buildings and facilities, habitats, hydrography, and many more essential data layers.

Increasingly, these information products are finding their way online as maps, comprehensive data layers, and interesting analytical models. This data comes to life for everyone as a living atlas, a collection of beautiful basemaps, imagery, and enabling geographic information, all of which are built into the ArcGIS platform. There, they are available for anyone to use, along with thousands of datasets and map services that have also been shared and registered in ArcGIS by users like you from around the world.

Web GIS is collaborative
Geography is the key, the web is the platform

Every day, millions of GIS users worldwide compile and build geographic data layers about topics critical to their work and for their particular areas of interest. The scope of information covers almost everything—rooms in a building, parcels of land, infrastructure, neighborhoods, local communities, regions, states, nations, the planet as a whole, and beyond, into other planetary systems. Web GIS operates at all scales, from the micro to the macro.

Geography is the organizing key; information in Web GIS is sorted by location. Because all these layers share this common key, any theme of data can be overlaid and analyzed in relation to all other layers that share the same geographic space.

This is a powerful notion that was well understood by mapmakers in the pre-digital era: tracing paper and later transparent plastic sheets were employed to painstakingly create "layer sandwiches" that could be visually analyzed. The desire to streamline this process using computers led to the early development of GIS. The practical term for this notion is "georeferencing," which means associating things using their locations in geographic space.

Now extend the idea—of georeferencing shared data—onto the web. Suddenly it's not just your own layers or the layers of your colleagues that are available to you, it's everything that anybody has ever published and shared about any particular

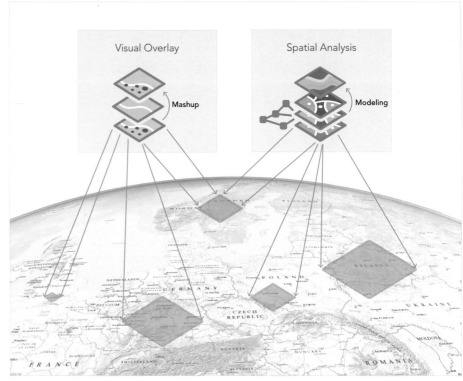

Visual overlay and spatial analysis can be applied to any patch of geography of any size on the planet.

geographic area. This is what makes Web GIS such an interesting and useful technology; you can integrate any of these different datasets from different data creators into your own view of the world, overlay them and then perform spatial analysis.

How GIS works
The science of geography

GIS is both a technology and a science. It relies on a simple notion of organizing data into discrete layers that are aligned (georeferenced) in relation to one another in geographic space.

Map Layers:
The secret advantage

Geographic datasets are presented in GIS as a series of dynamic, stacking map layers that cover a given extent (area). These layers can depict virtually any object (fixed or moving), boundary, event, or spatial phenomenon.

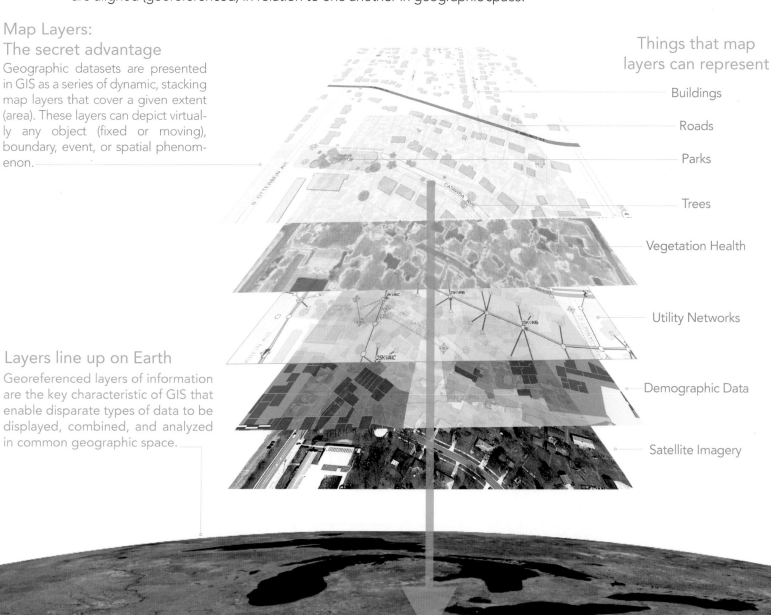

Things that map layers can represent

Buildings

Roads

Parks

Trees

Vegetation Health

Utility Networks

Demographic Data

Satellite Imagery

Layers line up on Earth

Georeferenced layers of information are the key characteristic of GIS that enable disparate types of data to be displayed, combined, and analyzed in common geographic space.

Westerville, Ohio, United States
40.1262° N, 82.9291° W

ArcGIS information items

Layers

You can think of the items eligible to be stored in ArcGIS as different types of geographic information. Now you can examine the three most primary and commonly accessed items: layers, web maps, and scenes.

Layers are logical collections of geographic data. Think about any map. It might contain such layers as streets, places of interest, parks, water bodies, or terrain. Layers are how geographic data is organized and combined to create maps and scenes; layers are also the basis for geographic analysis.

They can represent geographic features (points, lines, polygons, and 3D objects), imagery, surface elevation, cell-based grids, or virtually any data feed that has location (weather, stream gauges, traffic conditions, security cameras, tweets, and others). Here are a few examples of layers.

Nepal earthquake epicenters

Feature point data from in-ground data sensors.

Toronto traffic

Cell-based raster using historical predictive data.

Terrain of the Swiss Alps

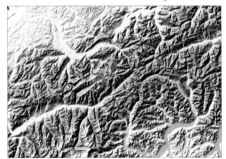

Tinted hillshade is a cell-based raster derived from an elevation surface.

Montreal, Canada, buildings

This 3D scene highlights layers for Montreal, Canada.

Sioux Falls parcels

Feature polygon data from cadastral surveys.

New South Wales wildfire tweets

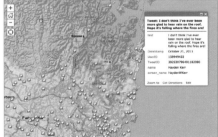

Feature point layer of tweets during 2013 New South Wales fires at #SydneyFires.

Web maps and scenes

Web maps

Web maps are the primary user interfaces by which work is done using ArcGIS. They contain the payload for GIS applications and are the key delivery mechanisms used to share geographically referenced information on the ArcGIS platform. Every GIS map contains a basemap (the canvas), plus the set of data layers you want to work with. If it's 2D, it's called a "web map." These are examples of two-dimensional web maps.

World earthquakes

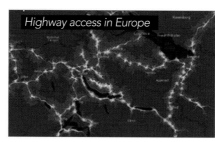

Highway access in Europe

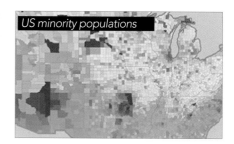

US minority populations

Scenes

The 3D counterpart to a web map is a scene. Scenes are similar to web maps (they combine basemap layers with your operational overlays), but scenes bring in the third dimension, the z-axis, which provides additional insight to study certain phenomena. These are examples of scenes.

3D integrated mesh

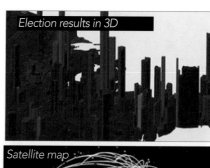

Election results in 3D

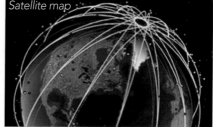

Satellite map

Geospatial analysis yields insights

GIS analysis is the process of modeling spatially, deriving results by computer processing, and then examining and interpreting those model results. Spatial analysis is useful for evaluating suitability and capability, estimating and predicting, interpreting and understanding, and much more.

ArcGIS provides a large set of modeling functions that produce analytical results. These functions typically generate new data layers and associated tabular information, enabling you to use ArcGIS to model just about any kind of spatial problem you can think of. (Chapter 5 delves into this area of ArcGIS in more detail.)

Sometimes analysis functions are built into the system. In many other situations, experienced users create their own models as analysis tools that can be shared with other ArcGIS users. These models can also be used to create new geoprocessing tasks in ArcGIS Enterprise (an advanced GIS server). So advanced users can create sophisticated analytical models that can be shared and accessed by other users who can work with their results.

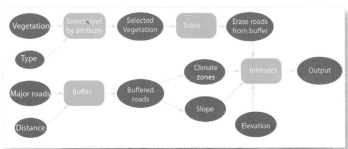

This shows a GIS workflow and resulting map used to model cougar habitat in the mountains and wildlands near Los Angeles.

Even beginners can apply spatial analysis. Practice and experience will help you expand the level of sophistication of your spatial analysis and modeling. The good news is that you can begin applying spatial analysis right away. The ultimate goal is to learn how to solve problems spatially using GIS.

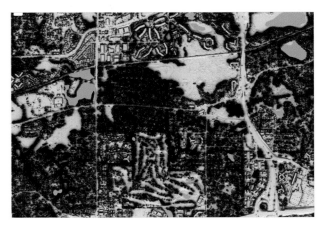

Minnesota models solar potential for the whole state by deriving expected solar radiation from critical raster and imagery layers. This enables citizens to perform a quick, high-level assessment of where solar power might be a practical alternative.

Apps extend the reach of GIS

Every GIS map has an interface—a user experience for putting that map to use. These experiences are called "apps," and they bring GIS to life for all kinds of people. And like other apps, these work virtually everywhere: on your mobile phones, tablets, in web browsers, and on desktops.

You'll learn much more about apps, these lightweight map-centric computer programs, in chapter 7, but for now you should know that as a publisher in ArcGIS, you can configure an app for specific users you want to reach by including a certain map or scene, and data layers, and setting other app properties. These configured apps are what you can save and share with selected users. And you can manage these apps as items in your ArcGIS account.

People have embraced the concept of apps. They get it. Many already use and value basic personal navigation maps, but expectations have been raised, and people increasingly want map apps that "do more."

The result is that map-based apps are the way that organizations extend the reach of their GIS in significant ways.

GIS on mobile devices is changing how we interact with geography. With your phone, you can access GIS maps and data anywhere, positioning you and your organization to leverage full GIS capabilities in the field. A GIS-enabled smartphone is also an advanced live data sensor.

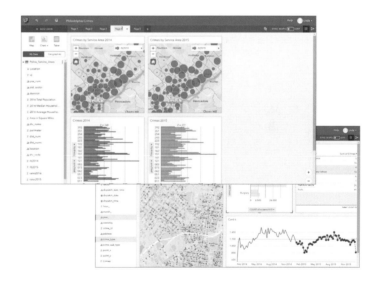

Insights is a new Esri app designed to explore and interpret GIS data layers using maps and charts. You can integrate additional rich layers to help investigate your data more deeply while discovering new patterns.

QuickStart

Connect to the ArcGIS platform online and on the desktop

Now it's time to get your hands on ArcGIS. If you're an existing user and already have an ArcGIS subscription (with Publisher privileges), as well as the ArcGIS Pro desktop application installed on your local machine, you're good to go and can skip to the next page. If you don't have these two things, read on.

▸ **Join the free Learn ArcGIS organization**

The majority of lessons in this book are carried out on the ArcGIS platform (in the cloud), and many require membership (with Publisher privileges) in an ArcGIS organization. Think of this as a data-rich learning sandbox available for students and others to practice with ArcGIS. With your membership, you can immediately begin to use maps, explore data, and publish geographic information to the web. Go to the Learn ArcGIS Organization link (http://go.esri.com/LearnOrg) and click the Sign Up Now link to activate a 60-day membership.

▸ **Install ArcGIS Pro**

ArcGIS Pro is a desktop application that you download and install on your local computer. If you don't already have ArcGIS Pro, you can get a limited-time license when you join the Learn ArcGIS organization. Check the system requirements, and then use the download button below to install the software on your local machine.

ArcGIS Pro is a 64-bit Windows application. To see if your computer will run ArcGIS Pro, check the system requirements.

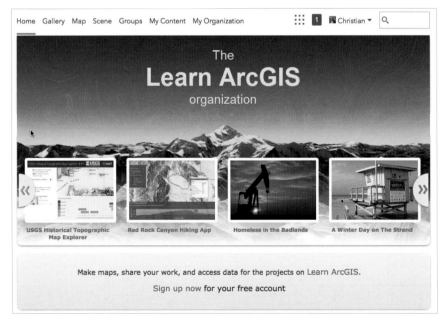

Download ArcGIS Pro

The Learn ArcGIS organization is set up specifically for educational use. It includes all the data required to complete the lessons found at Learn.arcgis.com. You can join this organization even if you already have another ArcGIS account.

Learn ArcGIS lesson

Demographic Analysis and Smart Mapping

The best way to get familiar with ArcGIS is to just dive in. In this Learn ArcGIS lesson you'll create a map of Detroit, Michigan, by adding and enriching a layer of ZIP Codes with demographic data from ArcGIS Online. You'll also apply smart mapping to style the layer, configure pop-ups to make the demographic information easier to read, and finally you'll report your findings by configuring a web app that tells a clear story based on your data.

▸ **Overview**

More children live in poverty in Detroit, Michigan, than in any other city in the United States. You work for a charity that supports community programs and poverty relief efforts. This year, the charity wants to direct its funding toward resources to help at-risk kids in Detroit. Your objective is to ensure the programs are offered where they're most needed.

▸ **Build skills in these areas:**

- Adding layers to a map
- Enriching layers with demographic data
- Styling layers with smart mapping
- Configuring pop-ups
- Editing item details
- Configuring a web app

Start Lesson

▸ **What you need:**

- Membership in the Learn ArcGIS student organization (or another organization with publisher privileges).
- Estimated time: 30 minutes to one hour

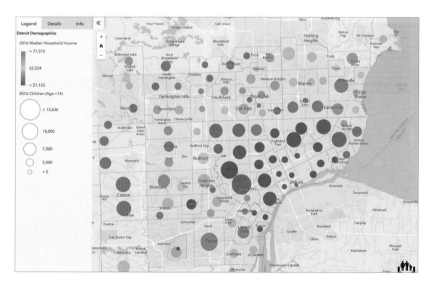

02

Mapping Is for Everyone
New ways to make, see, and use maps

GIS has radically changed how people create and engage with geographic information. Online maps shape the primary user experience, serving as both the means of creation and the mechanism for sharing and delivery. Interactive GIS maps are now used widely—traveling with us wherever we go. People have come to appreciate the power of combining layers of all kinds in their maps for a richer, more significant perspective about their world. With Web GIS, your maps can be accessed and put to work by virtually everyone, everywhere.

Online mapping is transforming GIS

Maps are important. Everyone understands and appreciates good maps. GIS users create and work with maps every day—they provide the basic experience and a practical interface for the application of GIS. Maps are also the primary way that GIS users share their work with others in their organizations and beyond. Maps provide a critical context because they are both analytical and artistic. They carry a universal appeal and offer a clarity and shape to the world. Maps enable you to explore your data and to discover and interpret patterns.

Online GIS maps can be created by anyone using Web GIS—and can be shared with virtually everyone. These maps bring GIS to life and can travel with us on our smartphones and tablets. Online maps have forever transformed computing and the web.

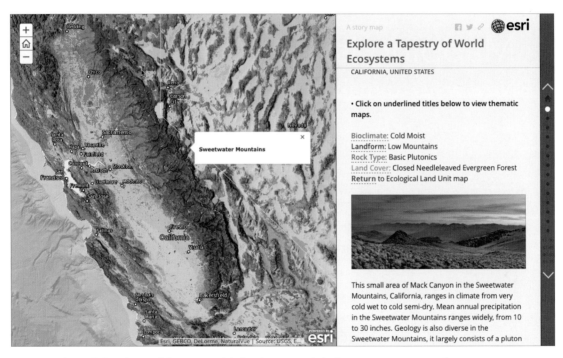

In 2015, the US Geological Survey published a new global ecosystem map of unprecedented detail. It is a mosaic of almost 4,000 unique types of ecological areas called ecological land units (ELUs) that are key to understanding the makeup of ecosystems.

Make no mistake, traditional printed maps are not going away. They continue to be important because they help you quickly grasp the broader context of a problem or situation. The best printed maps are true works of art that can stir your emotions and imagination. There's no comparable large-format document that communicates and organizes such large amounts of information so effectively and so beautifully. Cartographers using GIS will continue their craft of making astounding printed maps that teach and amaze.

And this will always be the case. Large-format printed maps and their digital cousins (such as PDFs) will continue to significantly occupy the good work of many mapping professionals. The difference now is that GIS tools have come of age for spectacularly high levels of professional cartography.

Meanwhile, a major online mapping revolution is under way, and the implications of this are far-reaching. We all know that consumer maps are ubiquitous on smartphones and the web. Map-based applications regularly rank among the most used programs on smartphones and mobile devices. Online maps have familiarized millions of people with how to work with maps, and this massive worldwide audience is ready to apply maps to their work in ever more imaginative ways using Web GIS.

GIS maps engage an audience for a purpose

Any map that you make can be saved and shared online—intended for a specific audience and expected uses. Online maps have an interface, a user experience (UX) that comes with each map, called an *app*. With the ArcGIS platform, a user (which you are by reading this book and becoming a member of the Learn ArcGIS organization) now has a wide range of options for designing and implementing purposeful maps and apps. The possibilities for engaging the audience that matters to you are endless.

This series of narrative maps tells the story of Americans of Japanese ancestry during World War II, acknowledged finally in 1982 by the US government as "... the bitter history of an original mistake, a failure of America's faith in its citizens' devotion to their country's cause and their right to liberty, when there was no evidence or proof of wrongdoing."

Mapmaking continues to evolve

Since the earliest recorded human history, maps have served to preserve and transmit geographic data by means of a visual representation. As such, they appeal to both the creative and logical aspects of our thinking; they're artistic but also logically organized around location.

The best maps unlock the full potential of the underlying data. Although the makers of the examples on this page weren't likely thinking about their knowledge of landownership or safe harbors as data, this information was valuable enough to commit to capturing in their respective maps. Although this book is primarily about modern digital interactive maps, it's practical and inspiring to recognize that good maps utilize information design principles that have been evolving for centuries.

This 2,500-year-old Babylonian clay map shows a plot of land in a river valley between two hills. Inscriptions record the owner of the 12-hectare parcel as a person named Azala.

This 1584 map of the waters off coastal Portugal contains a wealth of invaluable geographic information for mariners (such as sailing directions and the locations of harbors). The nautical atlas where it first appeared was the first of its kind, and an immediate commercial success. Modern sailors, while equipped with more accurate (and up-to-date) depictions, can still appreciate the critical value of the information, as well as the aesthetic beauty of its presentation.

Online maps can also provide the same powerful emotional and visual appeal of the great, printed maps. The age-old idea of information arranged spatially and thoughtfully presented for an intended audience will always be the guiding philosophy behind the work of geographic storytellers.

Successful maps work because they present some piece of geographic information in ways that illuminate, elevate, distinguish, intrigue, inspire, and promote fresh perspectives or points of view.

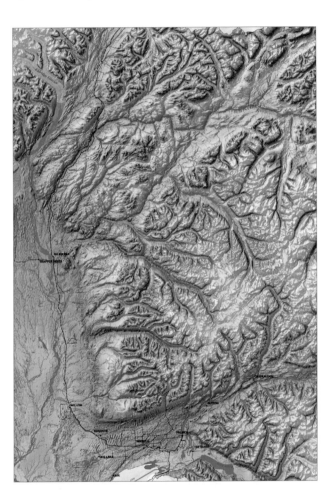

This map of daytime population patterns is effective because (A) it simply conveys several related datasets (predominance of workforce versus home daytime population), (B) it is multiscale with total US coverage, (C) it is interactive and can be "clicked through" to uncover more detailed information, and (D) it makes effective use of bright spot colors displayed on a muted but still readable basemap.

The Matanuska-Susitna Basin has a complex geography and dense hydrologic network of streams, rivers, and lakes. This map works by employing important techniques to present a striking view of the entire area.

Why GIS maps work

GIS maps work because they convey information about real things that matter to real people. All the examples featured here in some way manage to inspire, distinguish, promote, intrigue, or elevate their subject matter.

Thought leader: Scott Morehouse
Maps and geographic science enable a new kind of conversation

People are visual learners and seem to be instinctively attracted to maps. Maps help us instantly perceive patterns, relationships, and situations. They not only organize and present the rich content of our world, they offer a unique contextual framework for understanding, predicting, and designing the future.

GIS has a unique capability to integrate many kinds of data. It uses spatial location and digital map overlays to integrate and analyze the content of our world, uncovering relationships among all types of data. Maps and data form the underpinnings of GIS, which then organizes information into separate layers that can be visualized, analyzed, and combined to uncover meaning in data. This combination has resulted in a powerful analytic technology that is science-based, trusted, and easily communicated using maps and other forms of geographic visualization.

Online maps provide the user experience of working with and deriving answers from GIS. Maps provide windows into rich information—you can reach into a map to extract all kinds of related information. Maps also provide analytic functions that derive new information layers that enable us to answer whole new kinds of questions.

As you read this book, you'll understand that maps can be 2D and 3D, and they can animate information through time. Because you can add new layers from many sources, you can gain a new perspective and a deeper understanding about the problems and issues that you are trying to address.

Since 1981, Scott Morehouse has led the vision and organized the efforts to build commercial GIS software at Esri. He and his colleagues worked together to transform the set of early GIS concepts into numerous product releases, resulting in a great transformation in the role of mapping.

 Video: Modern GIS is transforming mapping

Perhaps the most profound role that maps play is that they provide a platform for engagement and conversations, for representing many points of view, for understanding the perspectives of others, and for helping humanity find answers to the many problems we face, things we care about— worthy goals that we can come together on.

The role of GIS maps

At their heart, web maps are simple

Web maps are online maps created with ArcGIS that provide a way to work and interact with geographic content organized as layers. They are shared across your organization and beyond on the web and across smartphones and tablets. Each map contains a reference basemap along with a set of additional data layers, plus tools that work on these layers. The tools can do simple things, such as open a pop-up window when you click on the map, or more complex things, such as performing spatial analysis and telling you the agricultural crop production in every county across the United States.

At their heart, GIS maps are simple. Start with a basemap and mash it up with your own data layers or those from other ArcGIS users. Then add tools that support what you want your users to do with your map: tell stories, perform analytical studies, collect data in the field, or monitor and manage operations.

Virtually anything you do with GIS can be shared using maps. And they can go anywhere. GIS maps work online and on any smartphone, and along with your supporting GIS work, they are accessible anytime.

Maps are how you deploy your GIS
A web map is easy to share with others. You simply provide a hyperlink to the web map you wish to share and embed it on websites or launch it using a wide range of GIS apps.

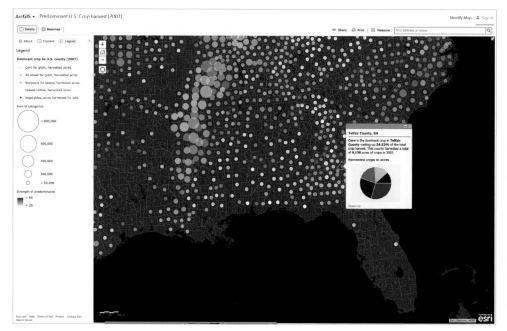

Maps are how users work with and apply ArcGIS, and can be used anywhere— in web browsers, on smartphones and tablets, and in desktop applications such as ArcGIS Pro. This web map is a window into a rich nationwide dataset about crops harvested county by county in the United States. Additional information products can be created using this map, including story maps and mobile-friendly apps.

Make and share a map

Five easy steps

Anyone can make, share, and use web maps. You can start by going through a short example. Suppose you want to make a map that allows you to explore the food, architecture, and design destinations for San Diego.

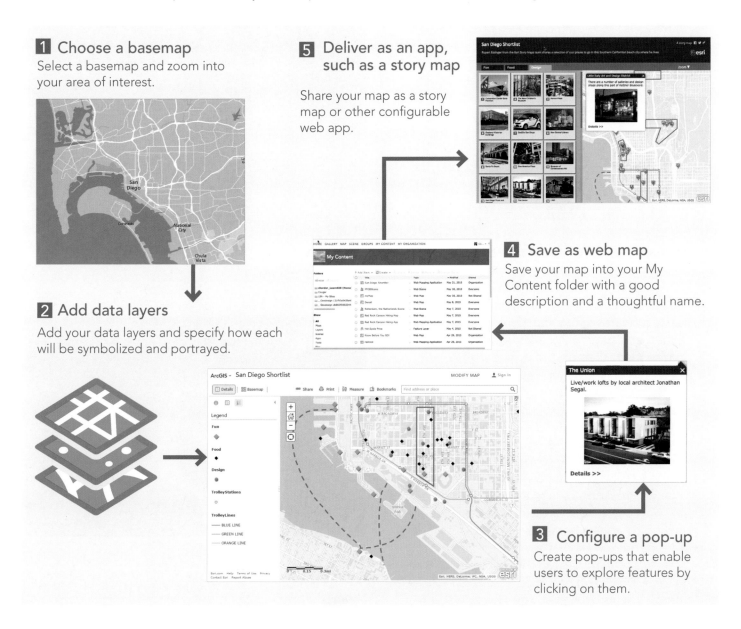

1 Choose a basemap

Select a basemap and zoom into your area of interest.

5 Deliver as an app, such as a story map

Share your map as a story map or other configurable web app.

2 Add data layers

Add your data layers and specify how each will be symbolized and portrayed.

4 Save as web map

Save your map into your My Content folder with a good description and a thoughtful name.

3 Configure a pop-up

Create pop-ups that enable users to explore features by clicking on them.

Basemaps and operational layers

The idea of a digital map mashup—recombining various geographic layers—is one of the great force multipliers in modern cartography. This ability to easily share and repurpose digital content has allowed individuals to create far more ambitious maps than would be possible if they had to work in isolation or start from scratch. The rise of the map mashup expanded cartography, so that anyone could build upon the work of others. Most of the thousands of maps created and shared every day within ArcGIS are created this way—maps that build upon the data, labor, and insights of the larger community. This era of collaborative GIS has empowered everyday citizens to participate in mapping as never before.

It starts with a basemap

In ArcGIS, map authors can readily access beautiful sets of professionally produced basemaps that provide the digital canvas on which to tell their stories. Each of the Esri basemaps has a theme or focus. Their range serves the need for almost any map type. Whether it's terrain, oceans, roads, or another of the many themes, the right basemap complements your subject and provides the background information critical to establishing its geographic context (locations, features, and labels). Each of the ArcGIS basemaps contains highly accurate and up-to-date information, at multiple zoom levels covering geographic scales from detailed building footprints to the entire planet. Providing data at each level of detail, for all locations on the globe, takes a small army of cartographers and eats up terabytes of data. The good news is that each of us can benefit immediately from those efforts. Some of the most widely used basemaps, such as those seen here, rack up billions of views every week.

ArcGIS includes a suite of basemaps that present the world in multiple cartographic styles. These three examples depict, from top, the Streets at Night theme, the Oceans basemap, and Imagery with Labels, the all-time most popular basemap.

Operational layers

Basemaps seem simple and relatively unobtrusive—and this is precisely their purpose. They should provide locational context (the "stage") for the content that is to be overlaid on them. Operational overlays carry the subject matter of the map and provide the purpose for making any map. A layer can be anything— emergency response incidents, your company's monthly sales, life expectancy, the location of oil and gas wells, or live traffic conditions. Merging a great basemap with one or more operational layers forms the heart of the modern online map.

The Living Atlas of the World

Some map authors are data creators interested in mapping their own data. Many other authors, however, need help finding operational layers; they know what they want to map but need guidance in finding the data to fully tell the story. Fortunately, ArcGIS provides access to an array of content to use in operational layers. The GIS community, including Esri, compiles and shares thousands of ready-to-use authoritative datasets in ArcGIS, covering everything from historical census data to environmental conditions derived from live sensor networks and stunning earth observations. And it's all in the Living Atlas of the World. You will learn more about it in chapter 4. Finding mappable, interesting geographic layers has never been easier.

Blending together ready-to-use basemaps, operational layers, and statistical graphs into a live, dynamic map allows you to share geographic content in a simple and concise format.

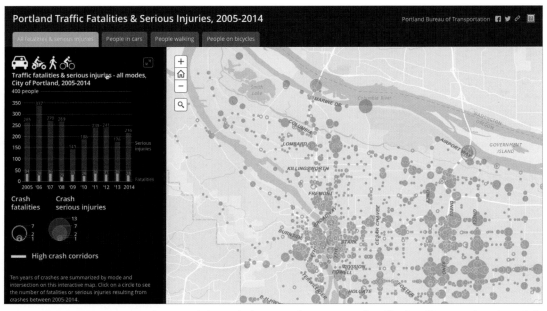

Just imagine trying to understand the subtle spatial patterns of traffic fatalities and serious injuries in Portland, Oregon, over a 10-year span by reading a spreadsheet. It would overwhelm anyone. By comparison, a map of that same data can be read and quickly understood with almost no training required. This is the power of web maps.

Web map properties

Continuous and multiscale

Web maps work across multiple scales. Zoom in to see additional details and gain insight. Online maps provide continuous pan and zoom. They literally have no edges—you can pan anywhere and zoom in for greater detail. Even if you don't have operational data for a particular area, the basemap will still provide reference.

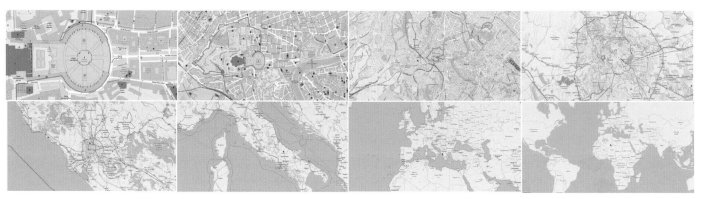

This web map contains the Open Street Map (OSM) basemap of the world. As you pan around and then zoom out to any spot on the planet, you will find levels of resolution and detail appropriate for that scale.

Pop-ups

Web maps are windows into a wealth of information. Click on a map location to "pop up" a report and explore the information behind it. Pop-ups help reach into the map for more detailed information that emerges on demand. A single window into a map can become a window into a world of related information, including charts, images, multimedia files, and analytics. The ability to link such a wide variety of content to the map has transformed how we think about maps. They've evolved from static containers of data to dynamic information vessels.

The best pop-ups don't have to be complex. In fact, the simple, informative ones, such as this story map about the World of Cheese, are effective because they deliver specifically what the map author wants you to know about that feature you clicked.

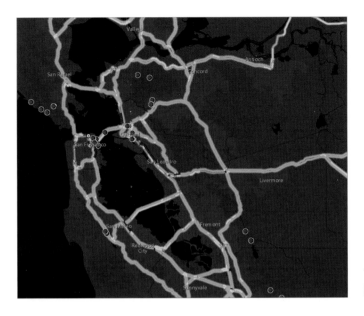

Real-time feeds

Your online maps are no longer static. They can be readily and immediately updated because your layers online can contain the latest, most accurate information. When your data changes, the maps that reference that layer are also updated.

This map features live-feed layers for traffic across the United States, Canada, and Mexico.

Mashup culture

Your maps can combine more than your own data. You can mash up your rich GIS data with information from other users—in fact, whatever is useful and relevant to your objectives from the entire world of GIS users.

This simple web map is a mashup of the locations of a well-known coffee company's San Francisco outlets.

Learning to smart map

The world is full of data, and maps help you make sense of it. There is a growing need to turn geographic data into compelling maps. People just want to create beautiful, interactive maps and infographics with live data, easily and with confidence. The smart mapping mission is to provide a kind of strong "cartographic artificial intelligence" that enables virtually anyone to visually analyze, create, and share professional-quality maps in just a few minutes, with minimal mapping knowledge or software skills.

Smart mapping is designed to give ArcGIS users the confidence and ability to quickly make maps that are visually pleasing and effective. Cartographic expertise is "baked" into ArcGIS, meaning it's part of the fundamental user experience of using ArcGIS. The map results that you see in front of you are driven by the nature of the data itself, the kind of map you want to create, and the kind of story you want to tell.

By taking much of the guesswork out of all the settings and choices that you could conceivably tweak, your initial map results are cartographically appropriate and visually pleasing. You can always change things at will, which you'll undoubtedly do as you gain more experience, but smart mapping gets you to something effective very quickly. You spend less time iterating and wrangling your maps into fulfilling your intention.

Spatial data exploration

One of the critically important capabilities of smart mapping is the added ability to interactively explore your data layers—for example, you can explore the range of values for median household income within each block group in your map by interacting with the histogram of median income values. The ability to interact with the data behind each map layer provides deeper insights into the questions you are trying to answer.

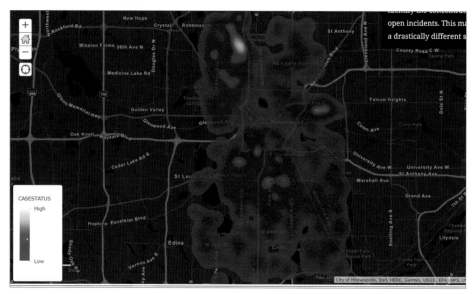

Smart mapping enables you to interact with and interpret the data behind your map. This quick guide introduces a fast, simple way to effectively analyze point data using heat maps within smart mapping.

Map design 101
Drawing your audience into the story you're telling

Maps are interactive, rewarding experiences, and not just pretty pictures. The most valuable maps are information products that are visually interesting from the first time you see them. Yet they reward you with additional information as you explore and interact with the map. When you touch the map, it responds by giving you details about the things you touch—touch a store and it tells you this year's sales to date with a chart of the previous three years' sales.

Great maps don't just happen automatically, though. You have to put a little bit of yourself into the effort, just like a great resume, which starts out as a template but requires your information—your data—as well as your interpretation to make it really sing. The data you are mapping won't tell its story without your help. Once you see the patterns emerging in the map, you can start emphasizing what's important, and de-emphasize everything else.

Try to always make "beautiful" maps. By that, we mean effective ones that are clear on first opening but that also engage users of all levels to drill in, explore, interrogate, and learn.

Start with the final result you have in mind and work backward. To paraphrase Roger Tomlinson, one of the fathers of GIS in the 1960s, you've got to know what you want to get out of a GIS to know what to put into it. Clicking aimlessly leads to no clear resolution. Have a clear idea of the story you want to tell about your data. Then get some test data and go for it.

 Make a Better Map: How to map counts and percents together

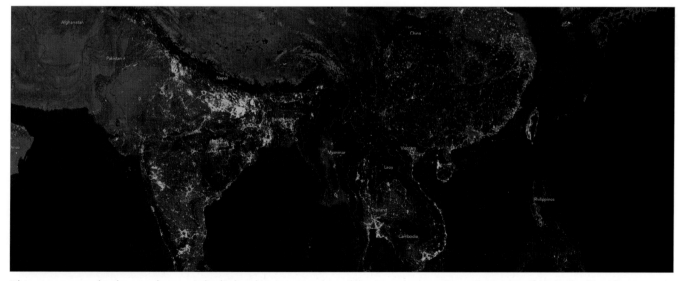

This story map looks at where night lights have turned on (blue) or dropped out (magenta) globally. Southeast Asia and the Indian subcontinent are shown here.

ArcGIS Pro: The cartographer's workhorse

ArcGIS Pro provides capabilities that enable serious mapmakers to create truly excellent maps, including support for highly sophisticated mapping workflows employed by professional cartographers. It includes tools for rich data compilation, for importing data from a multitude of publication formats, and for integrating this data with your own data to create consistent, accurate, and beautiful cartographic products for both printed and online maps. ArcGIS Pro is the workhorse application for serious cartographic production in both 2D and 3D and is used daily by hundreds of thousands of GIS users worldwide. This modern application builds on the tradition of great mapping with such enhancements as advanced 3D scenes.

The swisstopo map seen below, with its characteristic drawing and text style evolved by generations of Swiss cartographers over the past century, is widely considered the benchmark for 3D topographical mapping on paper. Today the agency uses tools like ArcGIS Pro to achieve the same results with computers.

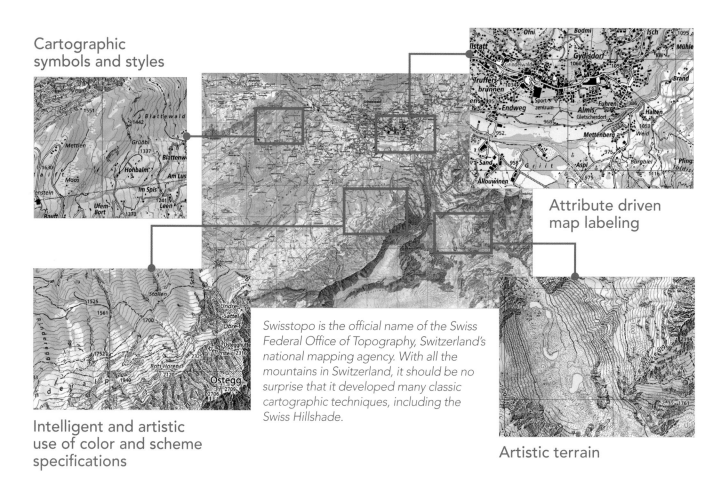

Cartographic symbols and styles

Attribute driven map labeling

Swisstopo is the official name of the Swiss Federal Office of Topography, Switzerland's national mapping agency. With all the mountains in Switzerland, it should be no surprise that it developed many classic cartographic techniques, including the Swiss Hillshade.

Intelligent and artistic use of color and scheme specifications

Artistic terrain

Maps into the third and fourth dimensions

While great cartography obviously predates the advent of computers, the digital era has yielded incredible fruit when it comes to mapping in the third (vertical) and fourth (temporal) dimensions. ArcGIS Pro provides tools that allow modern spatial storytellers to extend their maps into 3D and 4D (aka time).

Mapping a common climbing route from Mount Everest base camp to the summit at 8,848 meters.

Traffic patterns come alive using the animation tools in ArcGIS Pro.

Inspired by E.S. Glover's famous illustration, this 3D animation featuring the Los Angeles Basin in the late 1800s was created entirely in ArcGIS Pro.

QuickStart

Get inspired and learn current mapping techniques using curated selections of exemplary cartography at the Maps We Love website

What makes a good map? How can you engage people with a map? How do you make a map that offers unexpected insights and captivating appeal? We have been working on something at Esri that we hope will answer these questions: Maps We Love.

Maps We Love is an ongoing project where you will see the best of what's possible with ArcGIS. This is where you come for the inspiration, ideas, and information you need to turn your data into brilliant maps. We give you a behind-the-scenes look at important steps, plus resources (lots of links) so you can dig deeper into these topics. Maps We Love is designed to demystify mapmaking, to give you the confidence and assurance that you can make beautiful, effective maps.

Street cleanliness in Los Angeles.

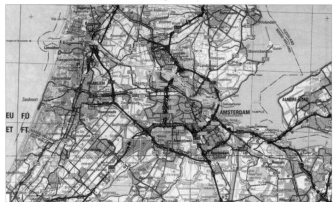

Two hundred years of Dutch cadastre.

Go to
Maps We Love

This guide, presented as a story map, explains what smart mapping is, and how to get started using this special capability.

Learn ArcGIS lesson

Cartographic Design with ArcGIS Pro

▶ Overview
In this lesson you'll use the US Department of Defense's recently released <u>Theater History of Operations data</u> to create an 11" x 17" wall map encompassing bombing and ground attack missions carried out by the United States from 1966 to 1974 during the Vietnam War.

You'll begin by downloading detailed country outlines from the Living Atlas. With over 3.1 million missions in the dataset, the challenge is to create a data visualization that is more of a macro view. Each mission is represented by a single, nearly transparent point, so that isolated missions are hardly visible, while areas of intensive saturation bombing are nearly opaque. Finally, you'll add a column chart of total missions by month and additional map elements including annotations, a scale bar, and title to complete the final product.

▶ Build skills in these areas:
- Accessing data from the Living Atlas
- Symbolizing and organizing data
- Creating charts
- Styling north arrows and scale bars

▶ What you need:
- ArcGIS Pro
- Estimated time: 1 to 2 hours

Start Lesson

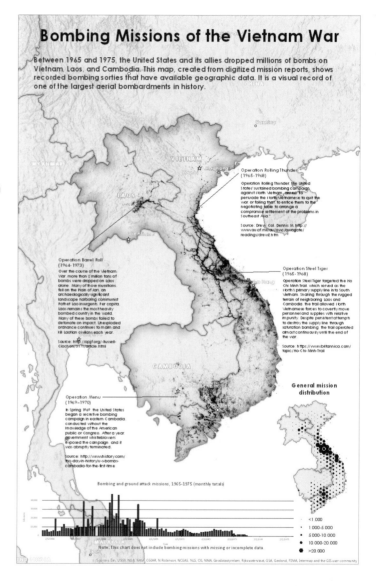

03

Tell Your Story Using a Map
Inform, engage, and inspire people with story maps

Everyone has a story to tell. Harness the power of maps to tell yours. Combine interactive maps and 3D scenes with narrative text and rich multimedia content to weave stories that get noticed.

Story maps
The fusion of maps and stories comes to life

Storytellers often turn to maps to illuminate and contextualize their words. Maps are the visual representation of where events happen. As such, maps and stories complement each other, but until recently they have existed more as side-by-side products and not as one integrated presentation. The big idea of this chapter is that narratives and geography can be combined into one experience: a story map.

Story maps use geography as a means of organizing and presenting information. They tell the story of a place, event, issue, trend, or pattern in a geographic context. They combine interactive maps with other rich content—text, photos, illustrations, video, and audio—within intuitive user experiences. While many story maps are designed for general, nontechnical audiences, some story maps can also serve highly specialized audiences. They use the tools of GIS, and often present the results of spatial analysis, but don't require their users to have any special knowledge or skills in GIS. This has resulted in a veritable explosion of story maps. (Go to storymaps.arcgis.com to see them come alive.) As you click through to the various story maps linked in this chapter or at the story maps website, take the freedom to immerse yourself in the various narratives. These are information products that reward exploration.

With today's cloud-based mapping platform, the fusion of maps and stories has finally come of age.

The International Union for Conservation of Nature's Red List of Threatened Species has evolved to become the world's most comprehensive information source on the global conservation status of animal, fungi, and plant species. This story map brings a geographic focus to their critical work, as well as an emotional focus by presenting beautiful paintings of the endangered and extinct species to a broad audience.

The world of story maps

A gallery of information-packed examples from around the globe

Every day the global Esri user community works to create the most authoritative scientific data on the world's most pressing and serious issues—much of it available for full-scale exploration on the ArcGIS platform. The imaginative uses of story maps and the live examples featured on these pages and in the Esri-curated Story Maps Gallery are designed to show the range of ways that such narratives can be used to convey rich and complex information.

This map, a collaboration between the Esri Story Maps team and PeaceTech Lab, uses crowdsourced data from Wikipedia to present a chronology of terrorist attacks around the globe. Wikipedia moderators include experts in the field of global conflict and terrorism, and the pages driving this story have been systematically revised numerous times since the beginning of 2016. As a result, the quality of the data on this map is constantly improving.

The International League of Conservation Photographers maintains this global image story map. Photographers share pictures and stories of how we are all connected through nature.

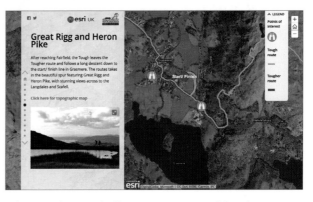

The Cumbrian Challenge is an annual fundraising event for wounded warriors that takes entrants over some of the most interesting and diverse landscape in England. This story map aids walkers and spectators alike.

Maps tell stories

What kinds of stories can you tell?

Describing places

Some maps do the very basic work of describing places. These are the maps we use to explore and navigate the world. Designed for intrepid travelers and armchair tourists, this entry from Norlisk, Russia, presents the story of deep cold.

Protecting lives

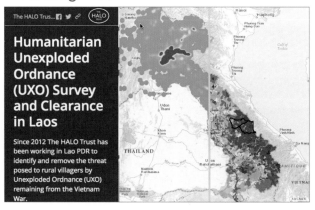

Vietnam remains plagued to this day by unexploded ordnance from the Vietnam War. This story map reveals how the locations of such ordnance continue to affect the people by denying safe access to agricultural land for rural villagers.

Revealing patterns

Although many might believe that America's largest fast-food purveyors are a nationally homogeneous bunch, this story map examines 94,000 locations to uncover the regional truth.

Presenting narratives

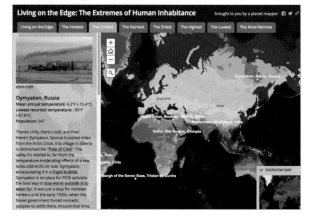

Take a tour of some of the most extreme inhabited regions of the earth, and learn about what it's like to live there. How do these places compare to where you live?

Visualizing the world

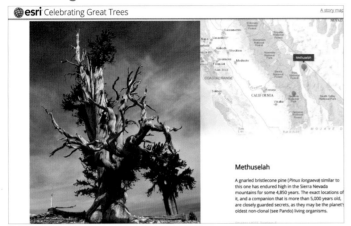

Our planet has countless millions of wonderful trees. On the occasion of Arbor Day, this map tells of a tiny proportion of trees which have gained fame as sacred or historic sites, or as specimens of unusual size, shape, or age—the botanical hall-of-famers.

Celebrating the world

In spite of their literally morbid function, graveyards and cemeteries are among some of the most fascinating locations on the planet, as playfully featured in this Halloween-themed story map.

Tracking the news

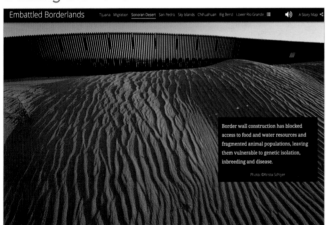

With US border security dominating the news cycles, this story map uses science and spatial thinking to objectively attempt to answer the question: Will the border wall strike a fatal blow to one of the most imperiled wild regions in North America?

Recounting history

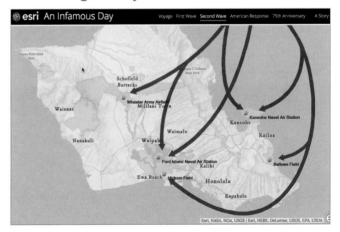

On December 7, 1941, the Japanese launched a surprise attack on Pearl Harbor. This exploration of the actual geography of the attack provides new context for an event that most people assumed they fully understood already.

The power of storytelling with maps

Storytelling carries the potential to effect change, influence opinion, create awareness, raise the alarm, and get out the news. Who authors story maps? Anybody—any individual or group with a desire to communicate effectively, including you. Here are a few examples, created by people just like you, to spur your imagination.

Literary research

California novelist Susan Straight created this rich story map to accompany a published essay. The map and essay explore the author's lifelong connection to regional American fiction, and conviction that these tales of "slaves and pioneers, indigenous and immigrants" can help us better understand the cultural differences that define the United States.

Historical education

San Francisco Fire Captain Henry Mitchell was on duty April 18, 1906, when the earthquake struck. For the next three days, he kept detailed notes, which were published over 100 years later by his grandson.

Scientific reporting

As climate change focuses the world's attention on the Arctic, new elevation data is being developed to improve the science. The data is brought to life and made accessible via this rich story map.

Thought leader: Allen Carroll
Why maps are so interesting

For most people, sight is the dominant sense, so when it comes to information delivery, most like it served visually. One way to think about it is to consider that as information publishers, we actually have relatively few ways to organize information. We can alphabetize it, but that's not very much fun. We can arrange it by time, chronologically, but that has its limitations. We can organize knowledge taxonomically by category or hierarchically in some kind of ranking. And then we come to spatial organization, the system that arranges things by where they are. This option offers unique insights and the potential to visualize information. Organizing by location is a particularly interesting and useful way to marshal information.

Another reason why so many relate to maps and geography is that we have no choice but to think and see spatially. We have to make sense of our surroundings and navigate through our world. Maps make sense of things. They lend order to complex environments, and they reveal patterns and relationships.

Maps can also be quite beautiful. They stimulate both sides of our brain: the right side that's intuitive and aesthetic, and the left side that's rational and analytical. Maps are this wonderful combination of both. It's this neat marriage of utility and beauty that I find so alluring.

For more than two decades, Allen Carroll told stories with maps at National Geographic. As the Society's chief cartographer, he participated in the creation of dozens of wall maps, atlases, globes, and cartographic websites. Today he leads the Esri Story Maps team, which uses state-of-the-art GIS technology, combined with digital media, to bring maps to life in new ways.

 Story Map Workshop from the 2016 Esri User Conference featuring Allen Carroll and Bern Szukalski

What kind of story do you want to tell?

Esri® Story Map Tour℠
A user experience for place-based narratives

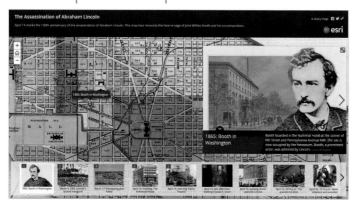

The _Story Map Tour_ app is ideal when you want to present a linear, place-based narrative featuring images or videos. Each "story point" in the narrative is geolocated. Users have the option of clicking sequentially through the narrative or browsing interactively.

Esri Story Map Journal℠
Create compelling multimedia narratives

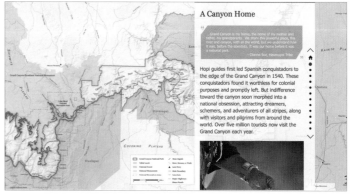

Designed for when you want to combine narrative text with maps and other embedded content, the _Map Journal_ app contains entries, or sections, that users simply scroll through to see associated maps, images, videos, illustrations, or web pages.

Esri Story Map Cascade℠
Create immersive scrolling narratives

The _Story Map Cascade_ app combines narrative text with maps, images, and other content in an engaging, full-screen scrolling experience. In a Cascade, sections containing text and in-line media can be mixed with "immersive" sections that fill the screen with maps and graphics.

Esri Story Map Series℠
Make it easy to browse a series of maps

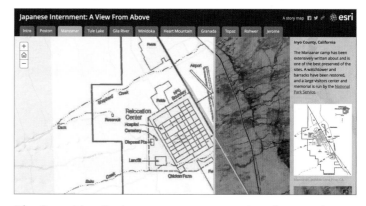

The _Story Map Series_ app presents a series of maps via tabs (shown above), numbered bullets, or an expandable "side accordion" control. In addition to maps, you can also include images, videos, and web content in your series to help tell your story and engage your audience.

Esri Story Map Crowdsource℠
Crowdsource your story map

The _Story Map Crowdsource_ app enables you to publish and manage a story that allows anyone to contribute photos with captions. Use it to engage a specific or general audience on any subject. Contributors can sign in with Facebook, Google, ArcGIS, or a guest account.

Esri Story Map Swipe℠ and Esri Story Map Spyglass℠
Compare two related maps

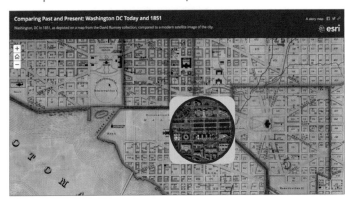

The _Story Map Swipe and Spyglass_ app enables users to interact with two web maps or two layers of a single web map. This allows you to present a single view, or to develop a narrative showing a series of locations or views of the same maps.

Esri Story Map Shortlist℠
A fun way to present points of interest

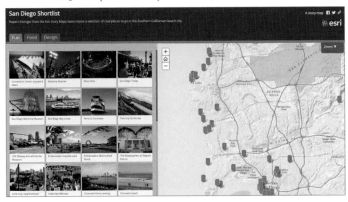

The _Story Map Shortlist_ app lets you organize points of interest into tabs that make it engaging for users to explore what's in an area. People can click on the places either on the tabs or in the map to find out more about them. The tabs automatically update as users navigate around the map.

Esri Story Map Basic
Let your map speak for itself

Story Map Basic puts all the emphasis on your map, so it works best when your map has great cartography and tells a clear story. You'll also want to take time to configure good-looking pop-ups, which can include text, images, graphs, videos, and more.

QuickStart
Things to consider when creating a story map

▶ Explore a story map about story maps!

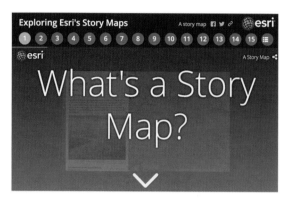

▶ **Think about your purpose and audience**
Your first step is to think about what you want to communicate with your story map and what your purpose or goal is in telling the story. Who is your audience? Are you aiming your story at the public at large, or a more focused audience, including stakeholders, supporters, or specialists who would be willing to explore and learn about something in more depth?

▶ **Spark your imagination**
Go to the Story Maps Gallery to see some examples handpicked to inspire you and that highlight creative approaches. Filter and search to check out how authors have handled subjects and information similar to your own. Explore. Get a good feel for what makes a good story.

▶ **Choose a story map app template**
Go to the Story Maps App to browse the app templates and choose the best one for your story map project. Each app lets you deliver a specific user experience to your audience.

▶ **Follow the instructions for the app template you choose**
See the Tutorial tab for the story map app template you choose for instructions on how to proceed. For example, here is the Tutorial for the Story Map Journal app template.

▶ **Share your story map and promote it**
When you've finished, you simply share your story map to launch it and make it go live. You can share it publicly or restrict it so it can be accessed only by people in your organization. To promote the story map to your audience, you can add links to it, embed it into your website, write a blog post about it, and share it on social media.

▶ **Subscribe to Planet Story Maps**
Get updates, tips and tricks, and other story map related news.

Get the latest Esri Story Maps news and tips delivered directly to you.

Planet Story Maps brings you the coolest new story map examples, tips, best practices, product announcements, news about contests and challenges, and more. It will inspire you and help you create better, more engaging story maps.

Are you ready to see the world through the new medium of story maps? And to tell your own stories?

Learn ArcGIS lesson

Tell the Story of Irish Public History

The Easter Rising, also known as the Easter Rebellion, was an unsuccessful armed insurrection in Ireland during Easter Week, April 1916. After the execution of 16 rebel leaders by the British administration, the Rising captured the political imagination and became an important rallying cry for separatist groups in Ireland. A 2006 commemorative event marked the first time that the civilian deaths of the rising were publicly and politically acknowledged.

▶ Overview

First, you'll use a comma-separated values (CSV) file to map the Rising's casualties. Then, you'll create a Public Information web app that includes social media posts about the Rising's commemoration. By comparing the locations of these social media posts to the locations of casualties, you'll probe questions about how the Rising has been perceived, remembered, and engaged with by the general public. In particular, you'll focus on how the story of civilian casualties has entered into patterns of commemoration and observations on social media. Lastly, you'll create a story map to share your findings in the context of some key locations during the Rising.

▶ Build skills in these areas:

- Adding layers from CSV files
- Creating a web app
- Creating a story map
- Drawing conclusions from maps and data

▶ What you need:

- Publisher or Administrator role in an ArcGIS organization
- Estimated time: 30 minutes to one hour

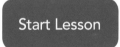

Start Lesson

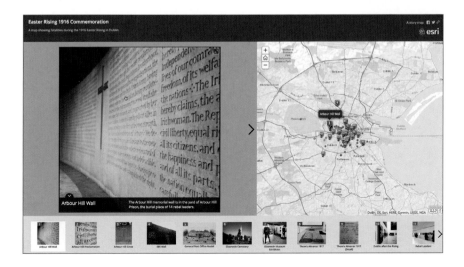

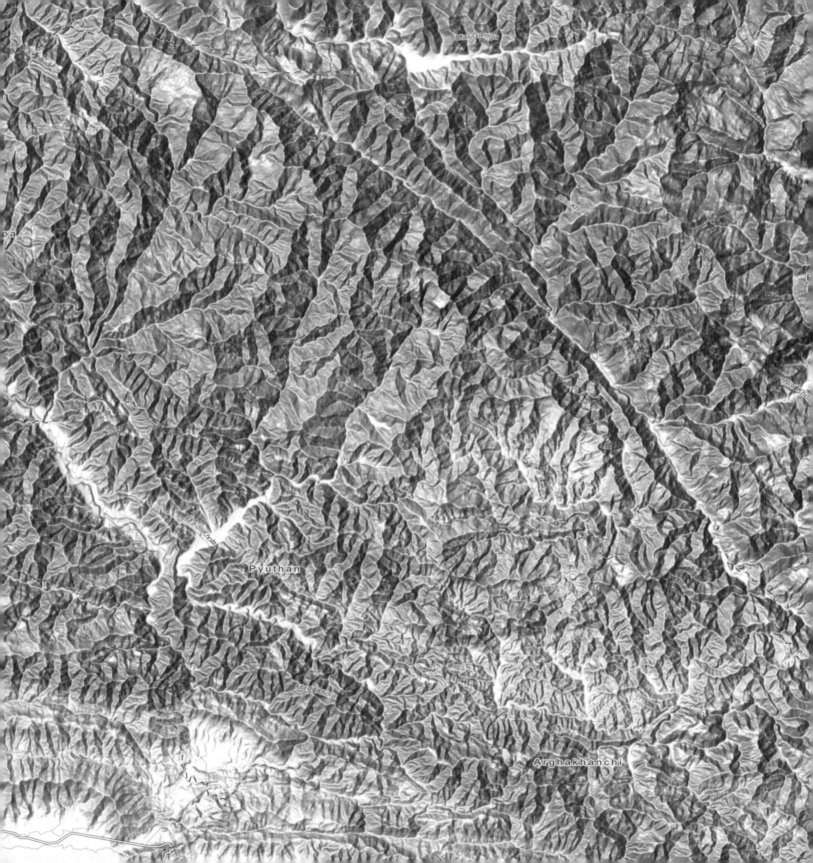

Great Maps Need Great Data
The Living Atlas of the World provides the foundation

ArcGIS Online is rapidly emerging as the platform of choice for the creation and dissemination of authoritative geographic data content. This Living Atlas of the World is a highly active network of contributors and curators whose output is accessed billions of times weekly. This chapter explains how this unique data ecosystem works, how to access its data, and how to contribute your own piece to the global GIS puzzle.

The Living Atlas of the World
The ArcGIS platform includes rich geospatial content

The Living Atlas of the World is a treasure trove of information, a dynamic collection of thousands of maps, datasets, images, tools, and apps produced by ArcGIS users worldwide (in conjunction with data curation and creation by Esri and its partners). It is the foremost and largest collection of global geographic information used to support critical decision-making. Think of it as a thematically organized, curated subset of the best available ArcGIS Online content, created and maintained by the GIS community. This deep and definitive catalog of information awaits your exploration and discovery. And that's the big idea of this chapter, that you can combine content from this repository with your own data to create powerful new maps and apps. And it's a two-way street: you can use the Contributor tools to add your own data to the Living Atlas.

The Living Atlas represents the collective work of the global mapping community—the people who use the ArcGIS platform as the system of record for their work. As such, it is fast emerging as the most extensive and authoritative source of geographic information on the planet.

Hunting down good data used to involve a lot of work just to get a GIS project started. These days, using ready-made basemaps and authenticated data from ArcGIS Online, GIS analysts are able to spend more time thinking analytically, which really gets to the heart of what makes global GIS work.

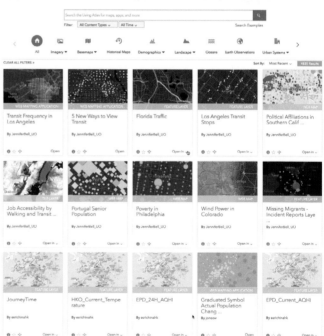

ArcGIS includes a Living Atlas of the World with beautiful and authoritative maps on thousands of topics. Explore maps and data from Esri and other organizations and combine them with your own data to create new maps and apps.

The GIS data community
A global network for creating and sharing

All GIS organizations have a core mission, a reason to exist, in support of their mandate and area of focus. As part of this work, these organizations are committed to building key authoritative data layers to support their core mission. This work includes the compilation of foundational data layers as well as standard basemap layers and operational data for their geographies and applications.

For such organizations—and they are innumerable in local, regional, state, and national levels around the globe—this information serves as the basis for all their comprehensive GIS applications. In the early days of GIS, the creation of these data layers from scratch was, in fact, job number one for these GIS organizations.

Foundation building
As accurate new *authoritative* geospatial data is developed, GIS users have been able to leverage their information resources in all kinds of GIS applications by extending their own work and helping their constituents.

The result is that all these different agencies have created data that is considered the official system of record maintained to support their mandated domain. The pace of migrating this data into Web GIS is increasing exponentially, and we now see many contributions coming online that fill in gaps for the entire world. The result is a continuous coverage of geographic information worldwide—the *GIS of the world*.

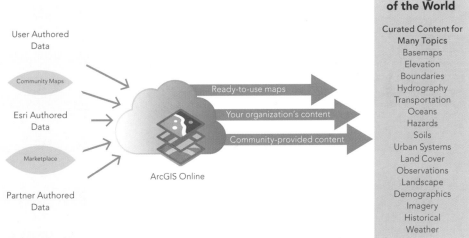

The Living Atlas of the World is a curated section of the larger ArcGIS Online data ecosystem, which includes user-, Esri-, and partner-authored data.

What kind of data is available?

Imagery

Imagery layers enable you to view recent, high-resolution imagery for most of the world; multispectral imagery of the planet updated daily; and near real-time imagery for parts of the world affected by major events, such as natural disasters.

Tokyo Station is in the center of the Japanese capital. This imagery layer is useful, and its metadata (resolution, age, and source) is but a click away.

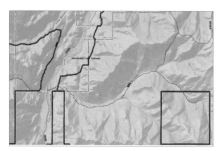

Boundaries and places

Many places are logically defined by a boundary. These map layers describe areas at many levels of geography, including countries, administrative areas, postal codes, census boundaries, and more.

Boundaries and Places are the bread and butter of vector GIS. Basically anything that can be depicted as point, line, or polygon features is found here. This web app features hiking trails in Idaho, a line feature.

Demographics and lifestyles

Demographics and lifestyles maps—of the United States and more than 120 other countries—include recent information about total population, family size, household income, spending, and much more.

This story map is a gateway to daytime population data for the entire United States.

Basemaps

Basemaps provide reference maps of the world and the context for your work. Built from the best available data from the GIS community of reliable data providers, these maps are presented in multiple cartographic styles provide the foundation for GIS apps.

This basemap provides a detailed representation of the world symbolized with a custom street map style that is designed for use at night or in other low-light environments. (Lower Manhattan is shown here.)

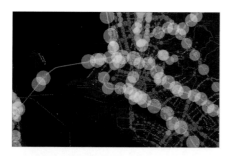

Transportation

These are the maps and layers that describe the systems that people use to move between places. They include a variety of global, national, and local maps on various topics from infrastructure projects to rest areas. Some of these layers are dynamic, such as the live World Traffic map, which is updated every few minutes with data on traffic incidents and congestion.

In places where high automobile and truck traffic exist, daily traffic counts reveal highly impacted patterns. (Los Angeles is shown here.)

Earth observations

These maps and layers are collected from sensors on the ground and in space. They describe our planet's current conditions, from earthquakes and fires to severe weather and hurricanes.

Whether you want to know how much snow fell in the Alps yesterday, or the current water temperature off the coast of Japan, these observations are available through ArcGIS.

Urban systems

These layers depict data about human activity in the built world and its economic activities and include such things as utility infrastructures, parcel boundaries, 3D cityscapes, housing, and employment statistics.

3D buildings, often in high detail, are being integrated into the Living Atlas. This is a scene of Rotterdam, the Netherlands.

Historical maps

This collection includes scanned raster maps and dynamic image layers. These layers can be viewed individually as a basemap or displayed against a current basemap for comparison purposes.

The David Rumsey Map Collection in ArcGIS Online features some of the most popular maps from the complete historical map collection, which focuses on rare 18th- and 19th-century North American and South American maps.

Basemaps
The setting for your story

A basemap provides a reference map for your world and a context for the content you want to display in a map. When you create a new map, you can choose which basemap you want to use. Change the basemap of your current map at any time by choosing from the basemap gallery or using your own basemap.

The evolution of basemaps has quietly changed the way of life for the everyday mapping professional. They make it easy to create most maps. Billions of ArcGIS maps utilizing these basemaps are created and shared every week. The examples seen here link to the detailed descriptions pages where each can be read about, and then opened in a live window.

These basemaps are multiscale, continuous, and provide global coverage:

Multiscale
This means that as you zoom into or out of a map, the features and detail that you see change. The ArcGIS basemap collection is continuous in scale. Zoom from the entire planet into the details of your neighborhood and down to a single parcel.

Global coverage
These maps cover the entire surface of the earth. Basemap coverage and levels of detail are improving each day as more data is added to the system.

Continuous
The extent of the map never stops; basemaps wrap around the surface of the earth.

World Imagery provides satellite and aerial, cloud-free imagery in natural color, at one meter or less, of many parts of the world and lower-resolution imagery worldwide.

With exactly the same imagery as the World Imagery basemap, this map includes political boundaries and place-names for reference purposes.

This comprehensive street map includes highways, major roads, minor roads, railways, water features, cities, parks, landmarks, building footprints, and administrative boundaries overlaid on shaded relief.

This basemap shows cities, water features, physiographic features, parks, landmarks, highways, roads, railways, airports, and administrative boundaries overlaid on land cover and shaded relief for added context.

Dark Gray Canvas

This dark basemap supports the overlay of brightly colored layers, creating a visually compelling map graphic that helps readers see the patterns intended by allowing your data to come to the foreground.

Light Gray Canvas

Like its dark counterpart, this basemap supports strong colors and labels against a neutral, informative backdrop. The canvas basemaps leave room for your operational layers to shine.

World Transportation

This reference map details the global transportation system with a street name reference overlay that is particularly useful on top of imagery.

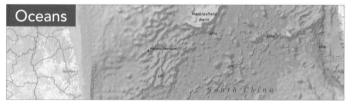

Oceans

The Oceans basemap (showing coastal regions and the ocean seafloor) is used by marine GIS professionals and as a reference map by others in the oceans and maritime community.

Terrain with Labels (Vector)

New in the fast-drawing vector format, this basemap features elevations as shaded relief, bathymetry, and coastal water features that provide a neutral background with political boundaries, and place-names for reference purposes.

OpenStreetMap

OpenStreetMap is the open collaborative project to create a free editable map of the world. Volunteers gather location data using GPS, local knowledge, and other free sources of information.

USA Topo Maps

This set of maps provides a useful basemap for a variety of applications, particularly in rural areas where topographic maps provide unique details and features from other basemaps.

USGS National Map

This composite topographic basemap of the United States by USGS includes contours, shaded relief, woodland and urban tint, along with vector layers, such as governmental unit boundaries, hydrography, structures, and transportation.

Demographics

This data about populations includes the basics, such as age and ethnicity, but also more sophisticated attributes such as people's wealth and health, their spending habits, and their politics. ArcGIS includes many hundreds of demographic variables (globally) that are accessible as maps, reports, and raw data that you can use to enrich your own maps.

The idea of data enrichment means that you can associate or append demographics to your local geography. This ability to combine your existing data with demographic variables specific to the problem being studied has opened a whole new avenue for everyone, not just consumer marketers, but epidemiologists, political scientists, sociologists, and any professional who wants to better understand a certain segment of the human population.

Demographers want to understand populations not only currently, but into the future. How will a given population group change over time? The art of forecasting current-year estimates on the basis of the decennial US Census, for example, is something that is carefully conducted by the demographic experts at Esri. One end product of this work is manifested as Tapestry Segmentation, which comes to life in the app below.

The amount of total demographic data available in ArcGIS Online is the epitome of "big data." The challenge for modern GIS developers and data scientists is to tame big data and deliver summarized, filtered, and interpreted information products. GIS really matters here.

Available US demographic data includes:

Updated Demographics
Accurate current-year estimates and five-year projections for US demographics, including households, income, and housing.

Census and American Community Survey
Census and American Community Survey (ACS) data used to analyze the impact of population changes on services and sites.

Tapestry Segmentation
Detailed descriptions of residential neighborhoods, including demographics, lifestyle data, and economic factors divided into 67 segments.

Consumer Spending
Data about products and services consumers are buying. Includes apparel, food and beverage, entertainment, and household goods and services.

Market Potential
Includes thousands of items that consumers want. The Market Potential Index measures consumer behaviors by area compared with the US average.

Retail Marketplace
Direct comparison between retail sales and consumer spending by industry. Measures the gap between supply and demand.

Business Data
Business Locations and Business Summary data from Dun & Bradstreet. Provides sales, employee information, and industry classification.

Major Shopping Centers
Statistics for thousands of major shopping centers, collected by the Directory of Major Malls. Includes name, total sales, and more.

Crime Indexes
Statistics about major categories of personal and property crime. Includes information about assault, burglary, and more.

Traffic Counts
Peak and low traffic volume of vehicles that cross a certain point or street location. Contains more than one million points.

Demographic and Statistics Atlas
This atlas shows how population is changing—growing in some parts of the United States, and shrinking in others.

Available global data includes:

Global Demographics
Recent demographics about total population, family size, household income, education, marital status, household type, unemployment, and more.

Global Spending
Total amount spent and amount spent per capita for categories such as food, clothing, household, medical, electronics, and more.

Open Data

Open Data allows organizations to use the ArcGIS platform to provide the public with open access to their geospatial data. Organizations use ArcGIS Online to create their own website and specify Open Data groups to share specific items. The general public can use Open Data sites to search by topic or location, download data in multiple formats, and view data on an interactive map and in a table. Here are some examples.

ArcGIS Open Data community

The ArcGIS Open Data community provides direct access to tens of thousands of open government datasets from thousands of organizations. These numbers are growing daily.

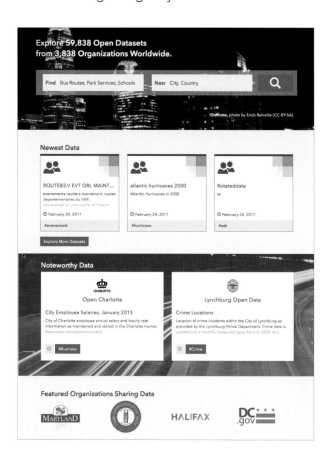

Maryland Open Data

Open Charlotte

Imagery

At the most basic level, imagery is simply pictures of the earth. Imagery can be immediate or taken across multiple time spans enabling us to measure and monitor change. Every image contains massive amounts of information and can be one of the quickest ways to collect data.

When it's integrated with GIS, imagery encompasses a broad collection of data about our world in the form of pictures from above—taken by satellites from space, aircraft flying over our cities, and collected by other sensors. Imagery represents the earth in digital pictures composed of millions of pixels. Satellite and aerial images are georeferenced pictures that overlay focused areas of our planet.

Because imagery sees the earth in unique ways, this enables us to both view and analyze our world using multiple perspectives. Depending on the satellite's sensors, imagery can provide access to both visible light as well as invisible aspects of the electromagnetic spectrum. This enables us to interpret what we can't see with the naked eye. We can visibly observe the presence or absence of water, classes of land cover and urbanization, the occurrence of certain minerals, human disturbance, vegetation health, changes in ice and water coverage, and a multitude of other factors. Imagery even enables us to automate the generation of 3D views of our planet.

Because the imagery collection is immediate, it enables us to monitor and measure change over time. For much more information about imagery and GIS, check out chapter 8 of this book and _The ArcGIS Imagery Book_.

Photographic

Aerial photography, historically on film, has gone digital. Still and video imagery from drones is on the rise. After this May 2014 Oklahoma tornado event, updated imagery for the scene appeared online within 24 hours.

Satellite

The use of satellite imagery to study the earth has seen a recent explosion thanks to the increase in the numbers of new imaging satellites and improvements to visualization and analysis software.

Multispectral

Electronic sensors in satellites and planes detect more than the human eye—information in the form of spectral bands. Once a band is captured as an image by a sensor, invisible bands can be displayed using the colors we see.

Landscapes

Landscape analysis layers

Landscape analysis underpins our efforts to plan land use, engage in natural resource management, and better understand our relationship with our environment. Esri has taken the best available data from many public data sources and provided the content in an easy-to-use GIS collection of datasets.

The map layers in this group provide information about natural systems, plants and animals, and the impacts and implications of human use of those resources that define the landscape of the United States and the rest of the world.

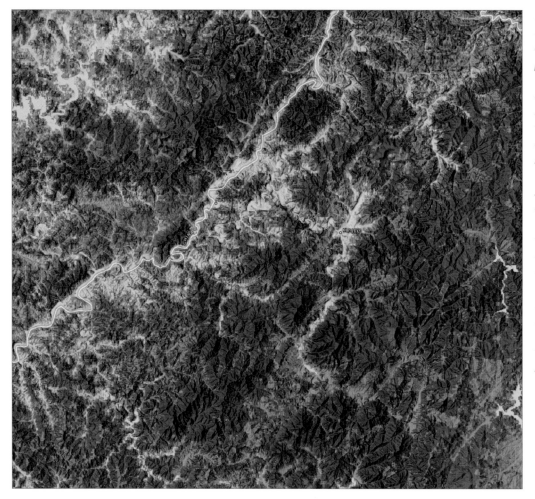

Ecological land units (ELUs) portray ecological and physiographic information about the earth. They provide an accounting framework for assessments of carbon storage and soil formation, and of important risk factors such as environmental degradation. ELUs also lend themselves to the study of ecological diversity, rarity, and evolutionary isolation. For example, we can identify the most diverse landscapes in terms of unique ecological land features. (North Korea shown here.)

Thought leader: Richard Saul Wurman
A map is a pattern made understandable

There's a notion that "the more you put on the map, the better the map," but there's a case where the opposite is true. Put two patterns together, and you'll discover a third. Pile on too much, and you can't discover a pattern at all.

The simple phrase "understanding precedes action" was an off-the-cuff remark I made that resonated as a truism, so the phrase stuck. Here's an example that illustrates why it's an important idea.

As cities incur more traffic, they add more freeways and highways. Yet does that actually solve the problem? Or does it spur the purchase of more cars, further crowding our freeways, while we consume fuel and generate more pollution? Adding more lanes only invites more traffic. The problem wasn't understood, but action was taken anyway.

Understanding precedes action. This is at the heart of the Urban Observatory, a longtime dream recently realized with the help of my friends at Esri. It's a simple idea. Yet simple is not necessarily reductive or dumbed down. In fact, it can be edifying. That's how I see it. And GIS is the key to this kingdom. It transforms mapping into a universal language and gives you the opportunity to ask questions and find answers visually. In fact, GIS allows us to ask better questions.

Richard Saul Wurman is an American architect and graphic designer. He has written and designed over 80 books and co-founded the TED (Technology, Entertainment, and Design) conference, as well as numerous other conferences. With a lifelong passion for creating understanding, Wurman has an extensive interest in maps, cartography, and design, culminating in his collaboration with Esri to create the Urban Observatory, which takes advantage of GIS as an integrative platform.

Explore the Urban Observatory

 Richard Saul Wurman and Jack Dangermond discuss the Urban Observatory

QuickStart

Using and contributing to the GIS data ecosystem

Browse the Living Atlas

ArcGIS includes a Living Atlas of the World with beautiful and authoritative maps and layers on hundreds of topics. These maps are shared by Esri, our partners, and members of the ArcGIS user community. The Living Atlas is curated so you, and others, can count on finding high-quality information for your ArcGIS apps.

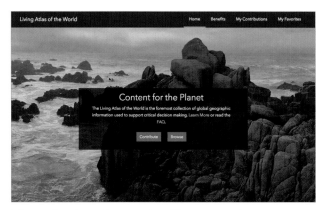

The collection of maps, intelligent map layers, imagery, tools, and apps built by ArcGIS users worldwide and by Esri and its partners is accessible at livingatlas.arcgis.com.

Contribute to the Living Atlas

You can help enrich the Living Atlas by sharing your maps and apps. To share your content items with the ArcGIS community, simply nominate your publicly shared ArcGIS Online maps and apps for review by our curators.

Unlock Earth's Secrets with Landsat Imagery

Landsat sees Earth in a unique way. It takes images of every location in the world to reveal hidden patterns in everything from volcanic activity to urban sprawl.

Community maps

Large-scale map layers add context and increased usability to basemaps. This story map details these map layers, features, and select contributor sites and provides examples of applied use.

Living Atlas Community Webinar

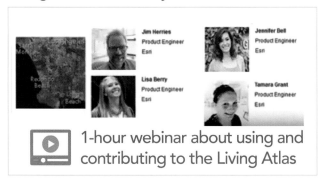

1-hour webinar about using and contributing to the Living Atlas

Learn ArcGIS lesson

Conduct a Demographic Analysis Using ArcGIS Demographic Data in Ottawa, Canada

You are the marketing manager for a chain of high-end beauty salons in the Ottawa, Canada, area. The owners recently opened a new location, and revenue isn't yet matching expectations— you need to create a new, loyal customer base! At a recent Women's Show convention, you held a drawing at your booth and collected a large amount of new contact information. You want to streamline this list to identify people who might realistically become new clients at the new salon, and send them a special promotional package to entice them to book an appointment.

To do this, you'll use ArcGIS Maps for Office to visualize salon locations on a map in Microsoft® Excel. You'll also add the locations of the new, potential customers and enrich the customer data with demographic information to ensure that you're pursuing more affluent customers. You'll analyze the data to determine which potential customers are within a reasonable driving distance from the new salon. Then you'll filter the analysis results to select only those particular customers so you can create a customized mailing list for your promotional package. Finally, you'll add a dynamic map slide to a Microsoft PowerPoint presentation to show your findings to your boss.

▸ Build skills in these areas:

- Creating a map from Excel data
- Styling the map to emphasize specific features
- Configuring pop-ups
- Enriching data using ArcGIS demographics
- Performing a drive-time analysis to find nearby features
- Filtering data to create a new map layer
- Sharing a map on ArcGIS
- Adding a dynamic map to a PowerPoint slide

▸ What you need:

- Publisher or Administrator role in an ArcGIS organization
- ArcGIS Maps for Office
- Approximate number of ArcGIS service credits: 20-25
- Estimated time: 30 minutes to one hour

Start Lesson

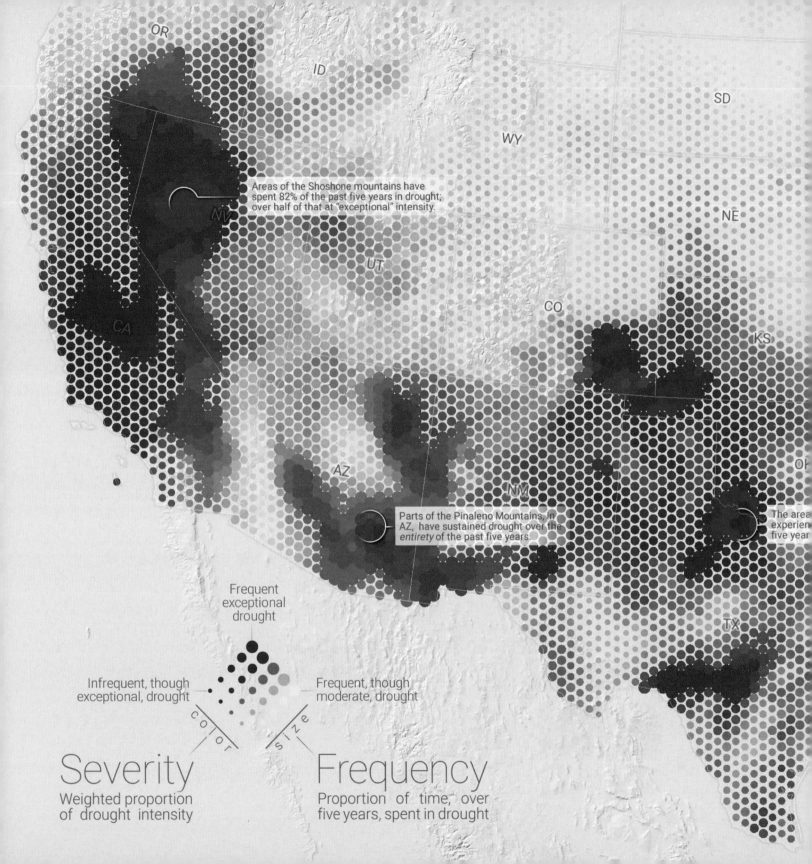

Areas of the Shoshone mountains have spent 82% of the past five years in drought; over half of that at "exceptional" intensity.

Parts of the Pinaleno Mountains, in AZ, have sustained drought over the *entirety* of the past five years.

The area experien five year

Frequent exceptional drought

Infrequent, though exceptional, drought

Frequent, though moderate, drought

color size

Severity
Weighted proportion of drought intensity

Frequency
Proportion of time, over five years, spent in drought

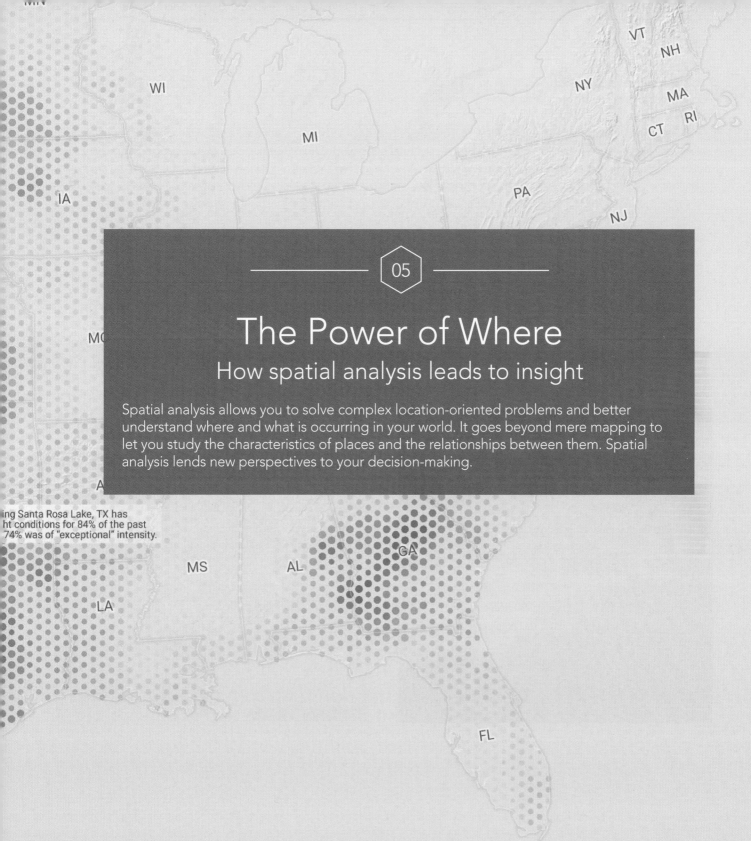

The Power of Where
How spatial analysis leads to insight

Spatial analysis allows you to solve complex location-oriented problems and better understand where and what is occurring in your world. It goes beyond mere mapping to let you study the characteristics of places and the relationships between them. Spatial analysis lends new perspectives to your decision-making.

ing Santa Rosa Lake, TX has
ht conditions for 84% of the past
74% was of "exceptional" intensity.

Geographic analysis

Have you ever looked at a map of crime in your city and tried to figure out what areas have high crime rates? Have you explored other types of information, such as school locations, parks, and demographics to try to determine the best location to buy a new home? Whenever we look at a map, we inherently start turning that map into information by analyzing its contents—finding patterns, assessing trends, or making decisions. This process is called "spatial analysis," and it's what our eyes and minds do naturally whenever we look at a map.

Spatial analysis is the most intriguing and remarkable aspect of GIS. Using spatial analysis, you can combine information from many independent sources and derive new sets of information (results) by applying a sophisticated set of spatial operators. This comprehensive collection of spatial analysis tools extends your ability to answer complex spatial questions. Statistical analysis can determine if the patterns that you see are significant. You can analyze various layers to calculate the suitability of a place for a particular activity. And by employing image analysis, you can detect change over time. These tools and many others, which are part of ArcGIS, enable you to address critically important questions and decisions that are beyond the scope of simple visual analysis. Here are some of the foundational spatial analyses and examples of how they are applied in the real world.

Determine relationships

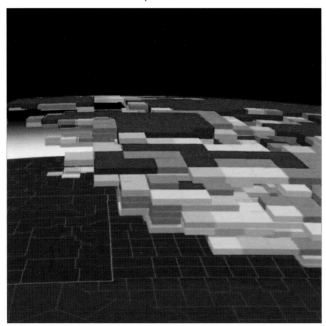

This 3D hot spot analysis of 20 years of storm cell data across the United States uses the vertical z-axis to represent time, so when tilted just right in a 3D viewer, it shows two decades of change in storm activity.

Understand and describe locations and events

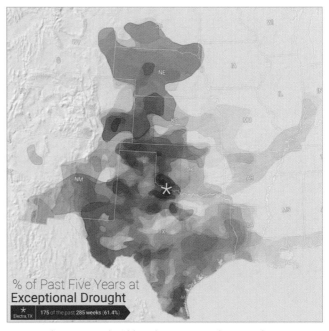

% of Past Five Years at
Exceptional Drought

Electra, TX 175 of the past 285 weeks (61.4%)

Using data complied by the National Drought Mitigation Center from numerous agencies, this map focuses on the widely varying degrees of drought in Texas from 2011 to 2016.

Detect and quantify patterns

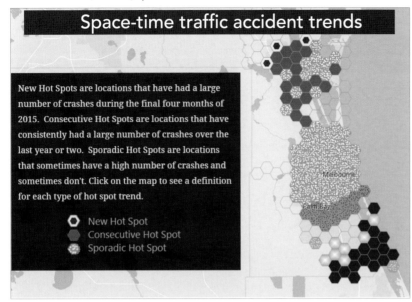

This space–time trend analysis of Florida auto crash data factors in time of day and underlying road conditions to identify new hot spots.

Make predictions

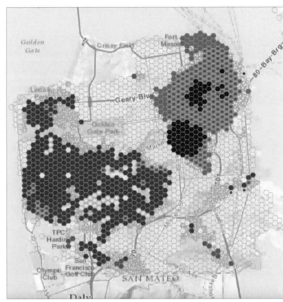

Statistical analyses can identify patterns in events that might otherwise seem random and unconnected, such as crimes in San Francisco.

Find best locations and paths

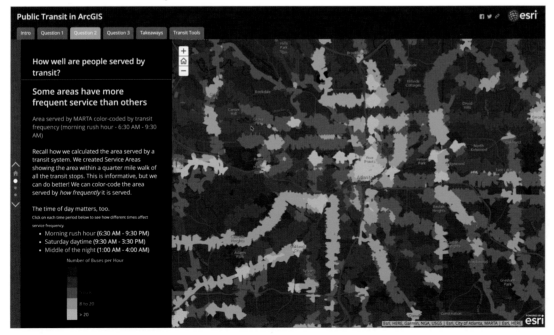

GIS analysis is used to explore how effectively the citizens of Atlanta are being served by public transit in this large urban community. Anyone who commutes understands that the time of day matters as well. You can use this story map to explore levels of transit service for different time windows.

How is spatial analysis used?

Pose questions, derive answers

Spatial analysis is used by people around the world to derive new information and make informed decisions. The organizations that use spatial analysis in their work are wide-ranging—local and state governments, national agencies, businesses of all kinds, utility companies, colleges and universities, NGOs—the list goes on. Here are just a few examples.

Crime studies

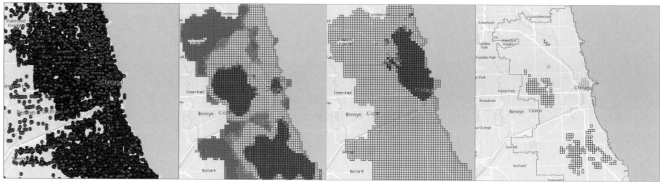

A spatial interaction model identifies the hot spots for crimes in Chicago.

Drought analysis

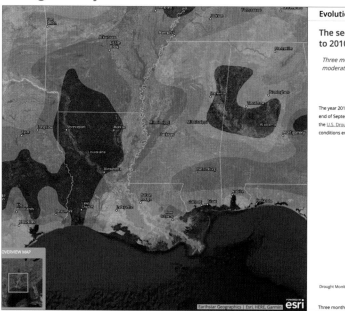

Evolution of the 2010-2015 Texas Drought

The seeds of drought are planted: A dry end to 2010

Three months of very little rain in the fall of 2010 set the stage for moderate drought to develop over 70% of Texas by the end of the year.

LEFT: Drought monitor map for December 28, 2010

The year 2010 began with a relatively wet winter, spring, and summer in Texas. At the end of September only 2% of the state was classified as being in drought, according to the U.S. Drought Monitor. By October, the beginning of the West's water year, dry conditions emerged.

December 28, 2010
(Released Thursday, Dec. 30, 2010)
Valid 7 a.m. EST

Drought Conditions (Percent Area)

	None	D0	D1	D2	D3	D4
Current	59.05	17.20	14.69	6.70	2.36	0.00
Last Week 12/21/2010	58.48	18.18	14.95	6.43	1.95	0.00
3 Months Ago 9/28/2010	60.05	26.79	10.07	2.79	0.30	0.00
Start of Calendar Year 12/29/2009	72.07	15.52	8.16	4.06	0.19	0.00
Start of Water Year 9/28/2010	60.05	26.79	10.07	2.79	0.30	0.00
One Year Ago 12/29/2009	72.07	15.52	8.16	4.06	0.19	0.00

Intensity:

D0 Abnormally Dry	D3 Extreme Drought
D1 Moderate Drought	D4 Exceptional Drought
D2 Severe Drought	

Drought Monitor Categorical Statistics for the Continental United States (CONUS) from December 28, 2010. (Source: U.S. Drought Monitor)

Three months of dry weather provided Texas and surrounding states with one of the

This temporal analysis of the evolution of the 2010–2015 Texas drought applies both raster and vector analysis methods. The project succeeds because of the attention to the final information product: a story map.

Green infrastructure

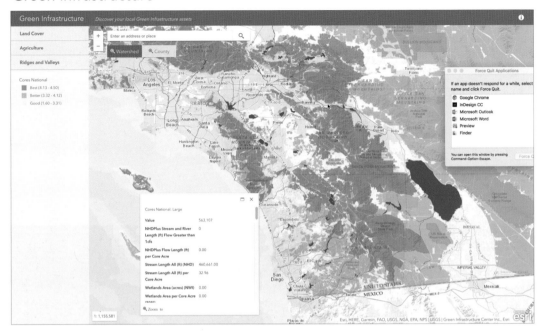

Esri's green infrastructure initiative set out to develop data for the continental United States of critical 100-acre patches depicting "intact habitat cores." It is making this data freely available as source data for land-use planning and to create information products that help everyone understand the importance of preserving the nation's remaining natural heritage.

Land-use planning

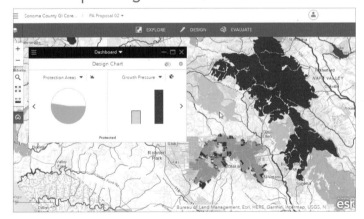

GeoPlanner℠ for ArcGIS® is a planning app used to evaluate opposing or competing land uses at local and regional scales. This screen capture shows a scenario where proposed protected areas (light green) are within areas of high projected population growth.

Automatic data interpretation

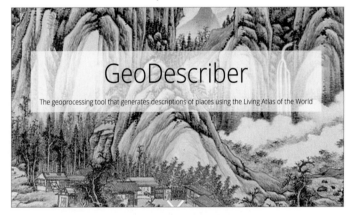

GeoDescriber analyzes landscape layers in the Living Atlas of the World to generate a short narrative of descriptive text to characterize the most important elements about a landscape.

Visualization

What can my map show me?

In many cases, just by making a map you are doing analysis. That's because you're making the map for a reason. You have a question you want the map to help answer: Where has disease ravaged trees? Which communities are in the path of a wildfire? Where are areas of high crime? It's also because when you make a map, as with any analysis, you're making decisions about which information to include and how to present that information. Effective visualization is valuable for communicating results and messages clearly in an engaging way.

Visual and visibility analysis

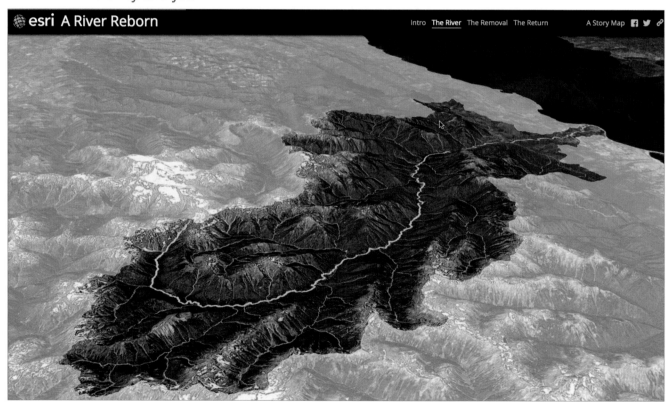

A surface displayed in 3D space has value as a visual display backdrop for draping data and analyzing it. This perspective scene shows a restored watershed and river draped on a digital elevation model of the terrain.

Visualizing solar radiation exposure

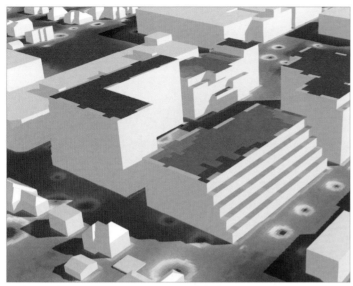

Solar radiation tools in ArcGIS enable you to map and analyze the potential for solar panels to generate electricity. (Naperville, Illinois, shown here.)

Assessing crop health

Multispectral imagery can provide a new perspective on crop health and vigor. The Normalized Difference Vegetation Index (NDVI) reveals healthy potato and canola crops in Saskatchewan, Canada.

Calculating viewshed

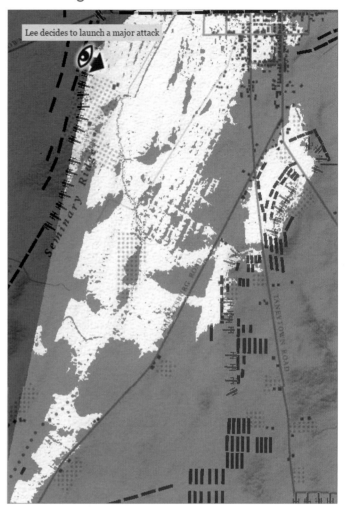

This historical story map uses GIS visibility analysis to tell the fateful tale of the Battle of Gettysburg in the American Civil War. At the moment General Robert E. Lee (at the red eye) committed to engage with Union troops, he could see only the troops in the light areas; everything shaded (the much greater part of the Union's strength) was invisible to him at that moment. Historians using personal accounts, maps of the battle, and a basic elevation layer were able to unlock the mystery of why Lee may have committed to battle facing such poor odds.

Spatial data and spatial analysis

Most data and measurements can be associated with locations and, therefore, can be placed on the map. Using spatial data, you know both what is present and where it is. The real world can be represented as discrete data, stored by its exact geographic location (called "feature data"), or continuous data represented by regular grids (called "raster data"). Of course, the nature of what you're analyzing influences how it is best represented. The natural environment (elevation, temperature, precipitation) is often represented using raster grids, whereas the built environment (roads, buildings) and administrative data (countries, census areas) tends to be represented as vector data. Further information that describes what is at each location can be attached; this information is often referred to as "attributes."

In GIS each dataset is managed as a layer and can be graphically combined using analytical operators (called *overlay analysis*). By combining layers using operators and displays, GIS enables you to work with these layers to explore critically important questions and find answers to those questions.

In addition to locational and attribute information, spatial data inherently contains geometric and topological properties. Geometric properties include position and measurements, such as length, direction, area, and volume. Topological properties represent spatial relationships such as connectivity, inclusion, and adjacency. Using these spatial properties, you can ask even more types of questions of your data to gain deeper insights.

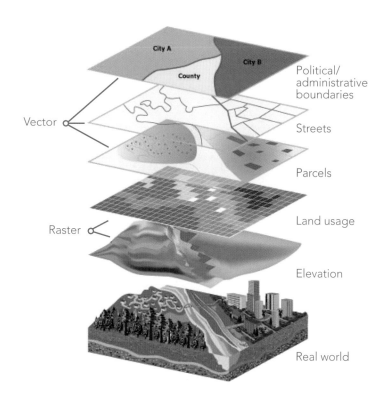

The idea of stacking layers containing different kinds of data and comparing them with each other on the basis of where things are located is the foundational concept of spatial analysis. The layers interlock in the sense that they are all georeferenced to true geographic space.

Anatomy of an overlay analysis

GIS analysis can be used to answer questions like: Where's the most suitable place for a housing development? A handful of seemingly unrelated factors—land cover, relative slope, distance to existing roads and streams, and soil composition—can each be modeled as layers, and then analyzed together using weighted overlay, a technique often credited to landscape architect Ian McHarg.

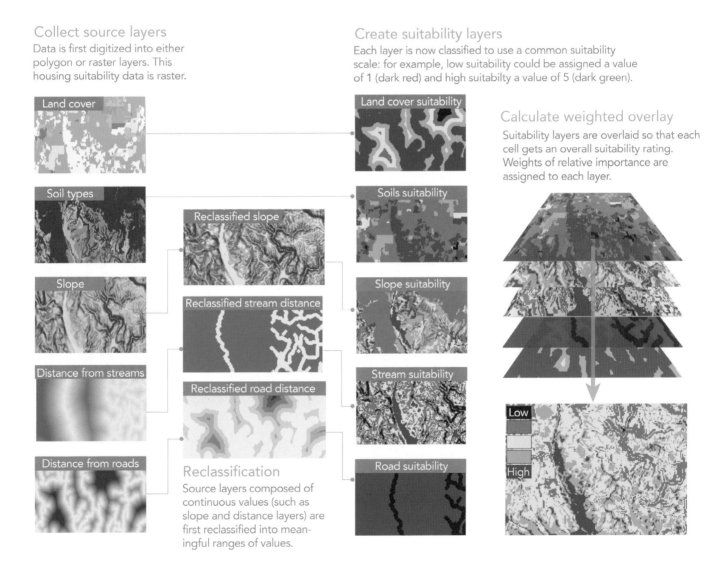

Collect source layers
Data is first digitized into either polygon or raster layers. This housing suitability data is raster.

Land cover

Soil types

Slope

Distance from streams

Distance from roads

Reclassified slope

Reclassified stream distance

Reclassified road distance

Reclassification
Source layers composed of continuous values (such as slope and distance layers) are first reclassified into meaningful ranges of values.

Create suitability layers
Each layer is now classified to use a common suitability scale: for example, low suitability could be assigned a value of 1 (dark red) and high suitabilty a value of 5 (dark green).

Land cover suitability

Soils suitability

Slope suitability

Stream suitability

Road suitability

Calculate weighted overlay
Suitability layers are overlaid so that each cell gets an overall suitability rating. Weights of relative importance are assigned to each layer.

Low

High

How to perform spatial analysis

The true power of GIS lies in the ability to perform analysis. Spatial analysis is a process in which you model problems geographically, derive results by computer processing, and then explore and examine those results. This type of analysis has proven to be highly effective for evaluating the geographic suitability of certain locations for specific purposes, estimating and predicting outcomes, interpreting and understanding change, detecting important patterns hidden in your information, and much more.

The big idea here is that you can begin applying spatial analysis right away even if you are new to GIS. The ultimate goal is to learn how to solve problems spatially. Several fundamental spatial analysis workflows form the heart of spatial analysis: *spatial data exploration*, *modeling with GIS tools*, and *spatial problem solving.*

Spatial data exploration

Spatial data exploration involves interacting with a collection of data and maps related to answering a specific question, which enables you to then visualize and explore geographic information and analytical results that pertain to the question. This allows you to extract knowledge and insights from the data.

Spatial data exploration involves working with interactive maps and related tables, charts, graphs, and multimedia. This integrates the geographic perspective with statistical information in the attributes. It's an iterative process of interactive exploration and visualization of maps and data.

Smart mapping is one of the key ways that data exploration is carried out in ArcGIS. It's interesting because it enables you to interact with the data in the context of the map symbology. Smart maps are built around data-driven workflows that generate intelligent data displays and effective default ways to view and interact with your information to see things such as your data's distribution.

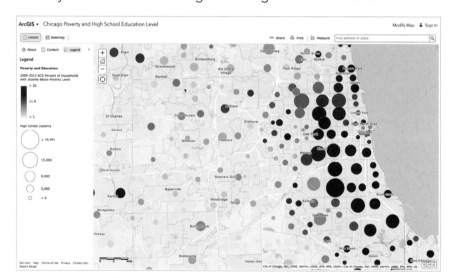

Smart mapping allows you to choose multiple attributes from your data, and visualize the patterns from each attribute within a single map using both color and size to differentiate (also referred to as bivariate mapping). This can be valuable for exploring your data, and allows you to tell a story using one map instead of many.

Combining interactive charts and graphs with GIS maps

Visualization with charts, graphs, and tables is a way to extend the exploration of your data, offering a fresh way to interpret analysis results and communicate findings. Typically you might begin by browsing through the raw data, looking at records in the table. Then maybe you'd plot (geocode) the points onto the map with different symbology and begin creating different types of charts (bar, line, scatter plot, and so on) to summarize the data in different ways (by district, by type, or by date).

Next, you can begin to examine the temporal trends in the data by plotting time on line charts. Information design is used to arrange different data visualizations to interpret analysis results. Combine a series of your strongest, clearest elements such as maps, charts, and text in a layout that you present and share.

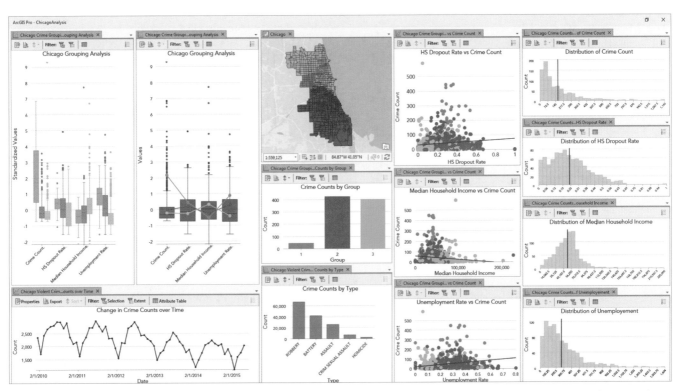

Finding the signal in the noise. Visualizing data through charts helps uncover patterns, trends, relationships, and structure in data that may otherwise be difficult to see as raw numbers. Depicting violent crime statistics from Chicago, a combination of chart and map styles work together to unlock patterns and meaning from what started out as pure tabular data.

Insights

Real-time exploration and analysis of maps and data

Insights for ArcGIS® is a browser-based analytic workbench that enables you to interactively explore and analyze your data coming from many sources. Insights enables you to quickly derive deeper understanding and powerful results through its rich, interactive user experience.

Insights for ArcGIS has the ability to integrate a variety of data sources for your analysis. It integrates and enables analysis of GIS data, enterprise data warehouses, big data, real-time data streams, and spreadsheets, and more. Insights for ArcGIS also leverages Esri's vast ecosystem of data, including the curated and authoritative Living Atlas of the World, by including a wider variety of information in analysis.

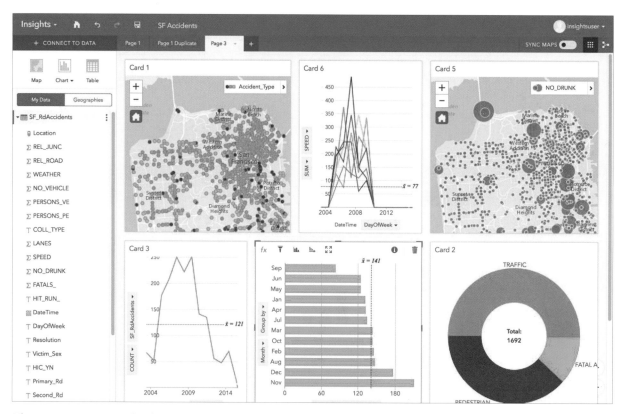

This screen capture displays crime incidents and uses descriptive statistics to summarize the human and financial costs of criminal activity in San Francisco over a five-year period. Click the image to view a video demonstration from the 2017 Esri Developer Summit.

The Insights workflow

1 Get started

Create an Insights workbook, visualize your data, and explore.

2 Add and manage data

Add data from different sources, and extend your data with location fields, attribute joins, and calculated fields.

3 Map and visualize

Create and interact with great-looking visualizations, thanks to smart defaults.

4 Find answers with spatial analytics

Update maps, draw buffers, use spatial filtering, and aggregate data across any geography and more.

Video demonstration:
Using Insights for ArcGIS to analyze global terrorist activity

Ten questions and answers:
Insights for ArcGIS

Modeling
Using the language of spatial analysis

Spatial analysis is the process of geographically modeling a problem or issue, deriving results by computer processing, and then examining and interpreting those model results. The spatial model that you create is based on a set of tools that apply operations on your data to create new results.

Data items and tools

Each geoprocessing tool performs a small yet essential operation on geographic data, such as adding a field to a table, creating buffer zones around features, computing the least-cost paths between multiple locations, or computing a weighted overlay to combine multiple layers into a single result.

A typical geoprocessing tool performs an operation on an ArcGIS dataset and produces new data as the result.

ArcGIS contains hundreds of analytical tools to perform just about any kind of analytical operation using any kind of geospatial information. For example, see the comprehensively rich set of operators found in the geoprocessing toolboxes that come with ArcGIS Pro. ArcGIS Pro also includes ModelBuilder, a visual programming application you can use to create, edit, and manage geoprocessing models.

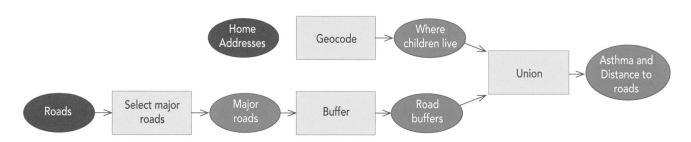

Here is an example of a spatial model created in ModelBuilder that enables the ability to explore potential relationships between the incidence of childhood asthma and air quality from heavy traffic.

Spatial analysis supports the automation of tasks by providing a rich set of tools that can be combined into a series of tools in a sequence of operations using models and scripts. Through spatial modeling, you can chain together a sequence of tools, feeding the output of one tool into another, enabling you to compose your own model.

Case study: *Puma concolor*
Modeling mountain lion habitat in Southern California

The metropolitan area of Greater Los Angeles region extends to 4,850 square miles (12,561 square kilometers) and represents the second-largest metropolitan area in the United States. The region has retained some of its original natural areas, and in the mountains surrounding the metropolis, the mountain lions (cougars) are the largest carnivores that live, hunt, and breed in this Southern California area. Our challenge is to ensure they survive. By connecting their remaining natural habitats to one another, in theory, this will allow the animals to seamlessly move between them.

This study analyzed ways to connect cougars located in several core areas with cougars in other geographically separated core areas. You will identify potential wildlife corridors that researchers and authorities can use to develop physical connections between cougar habitats located in the Santa Susana Mountains with habitats in the Santa Monica Mountains, the San Gabriel Mountains, and in the Los Padres National Forest. The complete workflow is described in the Learn ArcGIS lesson on page 84.

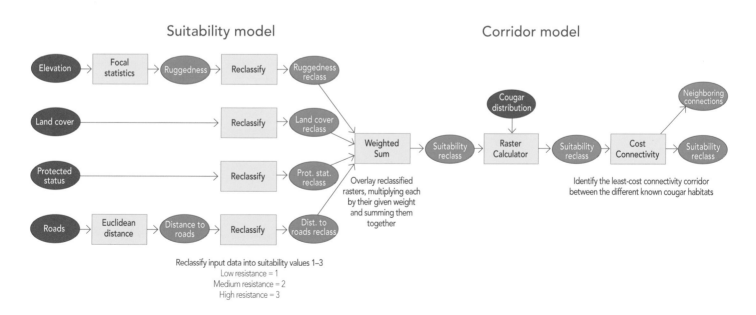

Suitability model

Corridor model

Reclassify input data into suitability values 1–3
Low resistance = 1
Medium resistance = 2
High resistance = 3

Overlay reclassified rasters, multiplying each by their given weight and summing them together

Identify the least-cost connectivity corridor between the different known cougar habitats

Spatial problem solving
A conceptual framework

Many types of problems and scenarios can be addressed by applying the spatial problem solving approach using ArcGIS. You can follow the five steps in this approach to create useful analytical models and use them in concert with spatial data exploration to address a whole array of problems and questions:

1. Ask and explore.
Set the goals for your analysis. Begin with a well-framed question that you'd like to address based on your understanding of the problem. Getting the question right is key to deriving meaningful results.

2. Model and compute.
Use geoprocessing to model and compute results that enable you to address the questions you pose. Choose the set of analysis tools that transform your data into new results. More often than not, you'll build a model that assembles multiple tools to model your scenarios, and then apply your model to compute and derive results that help you address your question.

3. Examine and interpret.
Use spatial data exploration workflows to examine, explore, and interpret your results using interactive maps, reports, charts, graphs, and information pop-ups. Seek explanations for the patterns you see and that help explain what the results mean. Effective exploration enables you to add your own perspectives and interpretations to your results.

4. Make decisions.
After exploring and interpreting your analytical results, make a decision and write up your conclusions and analytical results. Assess how adequately your results provide a useful answer to your original analysis question. Often new questions will arise that need to be addressed. These will frequently lead to further analysis.

5. Share results.
Identify the audience that will benefit from your findings and whom you want to influence. Then use maps, pop-ups, graphs, and charts that communicate your results efficiently and effectively. Share those results with others through web maps and apps that are geo-enriched to provide deeper explanations and support further inquiry. You can communicate your results using story maps as an effective way to share your findings with others.

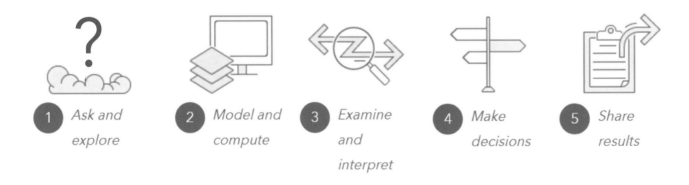

1 Ask and explore   2 Model and compute   3 Examine and interpret   4 Make decisions   5 Share results

Thought leader: Linda Beale

The challenge is making complex data understandable

Linda Beale is a geoanalyst and expert in spatial epidemiology—the examination of disease and its geographic variations. Her work contributes to the development of Esri's ArcGIS analysis and geoprocessing software, and as a Research Fellow in Health and GIS at the Imperial College London, she led the effort to publish The Environment and Health Atlas for England and Wales (*Oxford University Press, 2014*).

Geography plays a crucial role in health analysis. Fundamentally, it represents the context in which health risks occur; environmental hazards, risks, susceptibility, and health outcomes all vary spatially. Access to health care is characterized by both human and physical geographies. Furthermore, management and policy differ by location, and resources are allocated geographically. Health is important to everyone, but health analysis is challenging and demands a number of skills including epidemiology, statistics, and geographic information science. Spatial epidemiology is truly multidisciplinary, and although complex techniques are required for analysis, results must be accessible to everyone.

All these challenges were faced during the development of *The Environment and Health Atlas for England and Wales*. The atlas was developed with the ambitious goal of providing a resource—

for the public, for researchers, and anyone working in public health—with a collection of multiscale, interactive web maps that illustrate geographic distributions of disease risk and environmental agents at neighborhood scale.

Environmental monitoring and heath surveillance has advanced in recent decades, but emergencies continue to cause economic and social damage and, of course, loss of life. As the world becomes ever more interconnected both socially and economically, environmental and health impacts are felt at a wider scale than ever before. For example, following volcanic eruptions and nuclear accidents, or as a result of disease outbreaks such as avian influenza and Ebola, too often the impacts of environmental hazards fall disproportionately on the most vulnerable populations.

GIS offers the technology to explore, manipulate, analyze, and model data from multiple sources. With spatial analysis hazard mapping and predictions developed for risk assessment, you can use models to evaluate response strategies, and maps to illustrate preventative strategies and for risk communication and negotiation.

As technology has evolved, so have the science, the data, and the tools to test hypotheses and gain deeper insights into public health. We find ourselves at a time when, for many analyses, we are no longer awaiting technological or data advances. Instead, we should challenge ourselves to improve our understanding and public health through analysis.

QuickStart

ArcGIS spatial analysis tools are implemented in several places within the online and desktop environments.

▸ **ArcGIS Online**
The analytic capabilities of ArcGIS Online are accessed through the Analysis button on the map viewer:

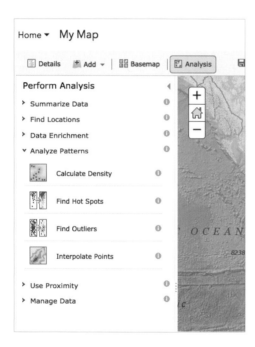

Credits: Some analysis tools consume ArcGIS Credits. Your Learn ArcGIS student account includes 200 credits.

▸ **Insights for ArcGIS**
At the time of this writing, Insights for ArcGIS requires ArcGIS® Enterprise. Look for its appearance in ArcGIS Online in the future.
.

▸ **ArcGIS Pro**
ArcGIS Pro is Esri's premier spatial analysis application. Its geoprocessing toolbox contains hundreds of spatial analytic tools. Your Learn ArcGIS Student membership allows complete use of the system for noncommercial purposes where you can learn spatial analysis by doing. Download the software; your license will be activated by the Learn ArcGIS organization.

▸ **Spatial Analysis MOOC**
This Massive Open Online Course (MOOC) runs periodically throughout the year. In this course you'll get free access to the full analytical capabilities of ArcGIS Online, Esri's cloud-based GIS platform.

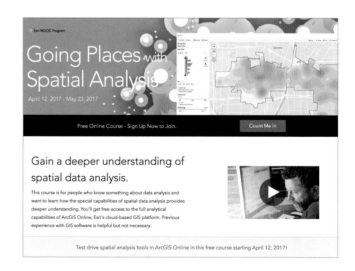

▸ **Online case studies**
An impressive set of spatial analysis case studies are on the ArcGIS Analytics website.

Learn ArcGIS lesson

Big Cats, Big Challenges—Claws, Cars, and Casualties

▶ Overview

The mega city of Los Angeles is one of the few world cities that have big cats living within the natural areas of the city. But the city landscape is becoming increasingly fragmented by urban development, roads, and freeways, thus leaving less space for mountain lions (cougars) to survive. As they attempt to cross roads and freeways in search of prey and mates, cougars often get struck and killed. For the Los Angeles cougar population to survive as well as maintain genetic diversity, and overall population health, a long-term solution must be found that will allow them to move safely between the isolated pockets of land they currently occupy.

In this project, your goal is to identify current cougar distributions and build a spatial model that identifies corridors we can establish to connect the various core mountain lion habitat areas within the city to each other.

The workflow emphasizes setting analysis goals which lead to questions that will give meaningful results. Following the workflow, you will examine and interpret analysis results, seek explanations for observed patterns, and explore their meaning from a spatial or temporal perspective. A strong emphasis of this workflow is on locating and using community and Living Atlas of the World data, and then contributing and sharing results and findings back to the community. The workflow also emphasizes the use of infographics and GeoEnrichment tools to provide deeper explanations and support for further inquiry.

▶ Build skills in these areas
- Converting and preparing data for analysis
- Creating and populating a geodatabase
- Adding and symbolizing data on a map
- Building a corridor analysis model using ModelBuilder
- Performing raster data classification and weighted overlay
- Generating a cost surface and least-cost path
- Sharing results with the public

▶ What you need
- ArcGIS Pro 1.4
- ArcGIS® Spatial Analyst extension
- Publisher or administrator role in an ArcGIS organization
- Estimated time: one to two hours

Start Lesson

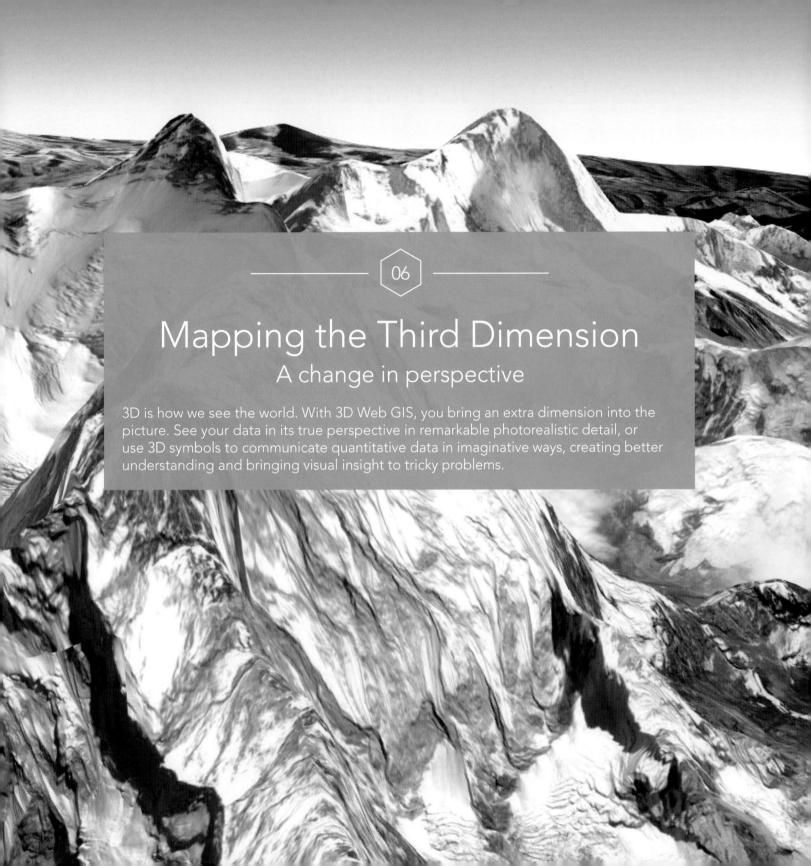

Mapping the Third Dimension
A change in perspective

3D is how we see the world. With 3D Web GIS, you bring an extra dimension into the picture. See your data in its true perspective in remarkable photorealistic detail, or use 3D symbols to communicate quantitative data in imaginative ways, creating better understanding and bringing visual insight to tricky problems.

The evolution of 3D mapping

Throughout history, geographic information has been authored and presented in the form of two-dimensional maps on the best available flat surface of the era—scrawled in the dirt, on animal skins and cave walls, hand-drawn on parchment, then onto mechanically printed paper, and finally onto computer screens in all their current shapes and sizes. Regardless of the delivery system, the result has been a consistently flat representation of the world. These 2D maps were (and still are) quite useful for many purposes, such as finding your way in an unfamiliar city or determining legal boundaries, but they're restricted by their top-down view of the world.

Three-dimensional depictions of geographic data have been around for centuries. Artistic bird's-eye views found popularity as a way to map cities and small-extent landscapes that regular people could intuitively understand. But because these were static and could not be used directly for measurement or analysis, they were often considered mere confections, or novelties, by serious cartographers, not a means of delivering authoritative content.

However, this is no longer the case since ArcGIS introduced the concept of a "scene," which is actually more than just a 3D map. In a scene, you can also control things like lighting, camera tilt, and angle of view. The mapmaker can craft a scene that creates a highly realistic representation of geographic information in three dimensions, which provides an entirely new way for the audience to interact with geographic content. Spatial information that is inherently 3D, such as the topography of the landscape, the built world, and even subsurface geology, can now be displayed not only intuitively and visually but also quantifiably and measurably, so that we can do real analysis and hard science using 3D data.

Badwater Basin

-279 feet | -85 meters

BADWATER BASIN

BELOW SEA LEVEL

(GFDL)

Three valleys over and only 85 miles eastward is North America's lowest point. The sign above is actually some distance from the actual lowest spot; salty crust conceals mud beneath, making for hazardous walking.

Some stories lend themselves well to 3D storytelling. Peaks and Valleys is a three-dimensional tour of our planet's highest and lowest spots.

Advantages of 3D

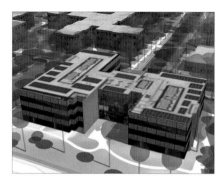

Vertical information

The most obvious advantage of a scene is its ability to incorporate vertical (and thus *volumetric*) information—the surface elevation of mountains, the surrounding landscape, the shapes of buildings, or the flight paths of jetliners. It's the power of the Z.

A campus 3D viewer is the ideal way to navigate around large grounds and also within the building. This visualization of the Esri campus in Redlands, California, allows us to visualize the campus in 3D, see selected indoor points of interest (POIs), as well as route to your desired destination.

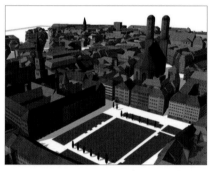

Intuitive symbology

In 3D, the extra dimension enables you to include more readily recognized symbols to make your maps more intuitive. You are able to see all the "data" from all viewpoints in situ. Every symbol that you recognize on a map saves you the effort of referring to the legend to make sure you understand what it shows.

No need for a legend on this 3D scene. The elements including the central plaza, the palm trees, and the structures are all instantly recognizable.

Showing real-world, bird's-eye views

Many of man's earliest maps, particularly of cities and smaller human habitations, were portrayed as scenes. These stylized maps were created as static 3D bird's-eye views and were successful in providing understanding of a place. Today's GIS authors interact with and see these scenes from many perspectives.

Explore redevelopment scenarios on the Portland riverfront.

Human-style navigation

For most of our living moments, we experience the world within a few feet of the ground. 3D allows us to replicate this view. With data presented from this approachable perspective, the size and relative positions of objects are intuitively understood as you wander virtually through the scene. There's no need to explain that you're in a forest or that a lake is blocking your route—the 3D perspective immediately makes the features recognizable.

Important 3D terminology
Getting the z-terminology straight

Maps and scenes

GIS content can be displayed in 2D or 3D views. There are a lot of similarities between the two modes. For example, both can contain GIS layers, both have spatial references, and both support GIS operations such as selection, analysis, and editing.

However, there are also many differences. At the layer level, telephone poles might be shown in a 2D map as brown circles, while the same content in a 3D scene could be shown as volumetric models—complete with cross members and even wires—that have been sized and rotated into place. At the scene level, there are properties that wouldn't make sense in a 2D map, such as the need for a ground surface mesh, the existence of an illumination source, and atmospheric effects such as fog.

In ArcGIS, we refer to 2D views as "maps" and 3D views as "scenes."

Local and global

3D content can be displayed within two different scene environments—a global world and a local (or plane) world. Global views are currently the more prevalent view type, where 3D content is displayed in a global coordinate system shown in the form of a sphere. A global canvas is well suited for data that extends across large distances and where curvature of the earth must be accounted for: for example, global airline traffic paths or shipping lanes.

Local views are like self-contained fish bowls, where scenes have a fixed extent in an enclosed space. They are better suited for small-extent data, such as a college campus or a mine site, and bring the additional benefit of supporting display in projected coordinate systems. Local views can also be effective for scientific data display, where the relative size of features is a more important display requirement than the physical location of the content on a spheroid.

This scene follows Magellan's voyage around the world. In geography and history classrooms, rich 3D user experiences help students understand the challenges and results of such a trip.

Accurately textured, realistic 3D building models for downtown Indianapolis, Indiana.

Surfaces

A "surface" is like a piece of skin pulled tight against the earth. Surface data by definition includes an x-, y-, and z-value for any point. A surface can represent a physical thing that exists in the real world, such as a mountain range, or it can be an imagined surface that might exist in the future, such as a road grading plan. It can even show a theme that only exists conceptually, such as a population density surface. Surfaces come in a wide variety of accuracies, ranging anywhere from high-resolution, 1-inch accuracy all the way down to a low-resolution surface with 90 meters or coarser accuracy.

Surfaces are fundamental building blocks for nearly every scene you will create because they provide a foundation for draping other content. Sometimes the surface itself is the star of the show (like a scene of Mount Everest). Other times the surface serves a more humble role of accommodating other crucial scene data, such as aerial imagery or administrative boundaries. And surfaces can also provide base-height information for 3D vector symbols, such as trees, buildings, and fire hydrants, for which their vertical position within the scene might not otherwise be known.

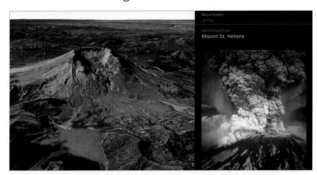

This scene presents surfaces of interesting places around the globe, featuring the World Imagery basemap along with 3D terrain layers. You can click on the slides in the scene to explore them and navigate the scene to see different perspectives for each place.

Real size and screen size

Symbolizing features using a real-world size is extremely common in 3D. For example, it's expected that buildings, trees, and light poles all be displayed at the same relative size in the virtual world as they exist in reality. Even some thematic symbols, such as a sphere showing the estimated illumination distance of one of the light poles, will help communicate the notion of real-world size.

However, it is also useful to have symbols in the scene that are an on-screen size instead. That is, as you zoom in and out of the scene, the symbol always displays with the same number of pixels on the screen. This effect is analogous to a 2D map layer whose symbol sizes do not change as you move between map scales.

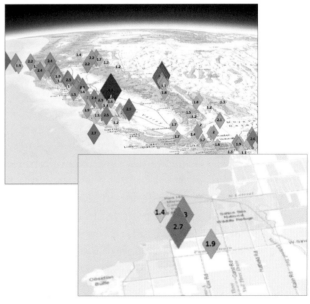

This earthquake map of California features screen-size symbols that remain the same size regardless of how far you zoom in and out, or where you pan to.

New worlds of 3D data
Point clouds, underground GIS, and more

3D data is increasingly available from a wide variety of different sources. The examples featured here hint at the possibilities. Take some time to click through these apps on your computer. These and many other innovative examples are collected in the ArcGIS Web Scenes gallery.

Lidar

Light detection and ranging (lidar) is an optical remote-sensing technique that uses laser light to densely sample the surface of the earth, producing highly accurate x, y, and z measurements. Lidar, primarily used in airborne laser mapping applications, is emerging as a cost-effective alternative to traditional surveying techniques such as photogrammetry. Lidar produces mass point cloud datasets that can be managed, visualized, analyzed, and shared using ArcGIS.

Scheidam, the Netherlands

Integrated mesh

Integrated mesh data is typically captured by an automated process for constructing 3D objects from large sets of overlapping imagery. The result integrates the original input image information as a textured mesh using a triangular interlaced structure. An integrated mesh can represent built and natural 3D features, such as building walls, trees, valleys, and cliffs, with realistic textures and includes elevation information. Integrated mesh scene layers are generally created for citywide 3D mapping and can be created using Drone2Map™ for ArcGIS®, which can then be shared to ArcGIS Desktop or web apps.

Marseilles, France

Drone imagery data

In the past few years, drones have become an increasingly common way to capture high-resolution imagery of local areas. Drone images are generally tagged with geographic information that describes where each image was taken, making them ready for use in ArcGIS. Drone2Map for ArcGIS not only allows you to view raw drone images on a map, but you can also create both 2D maps and 3D scenes from the images.

Calimesa, California

The world beneath our feet

By default, navigation below ground is disabled to avoid accidentally zooming under the ground surface of a 3D scene and becoming disoriented. If, however, your scene contains data that correctly belongs underground—such as subsurface utility pipes or geological bodies—you can enable this capability for the 3D scene.

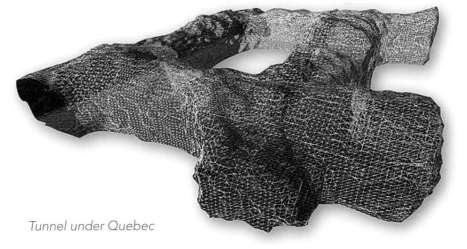

Tunnel under Quebec

Representing the world in 3D

Montreal, Canada

Pasadena, California

Photorealistic

Photorealistic views are essentially attempts to re-create reality by using photos to texture your features. These are by far the most common type of scene, with enormous amounts of effort put into making the virtual world look exactly as if you were there in person. Authors of this content create virtual worlds for simulation, for planning and design, and for promotional videos and movies. The specification remains very simple: *look out the window, and make the virtual world appear like that.*

In a GIS context, photorealistic views are extremely well suited for showing the public how a place has changed, or is expected to change, through time. That could mean what the cityscape will look like after a proposed building is constructed, or what a region looked like when dinosaurs roamed the earth. A photorealistic view takes the onus off users of imagining what the state of the world would look like, and simply shows them.

3D cartographic

Using 3D elements to represent data and other nonphotorealistic information is the next frontier. The idea is to take 2D thematic mapping techniques and move them into 3D. These maps are powerful, eye-catching, and immersive information products, often viewed as navigable scenes or packaged as video to control the user's experience and deliver maximum impact.

Philadelphia

Virtual reality

A 3D scene quickly starts to feel like virtual reality when photorealistic and thematic techniques are used in combination. The photorealistic parts of the scene provide a sense of familiarity to the user, and the thematic parts can convey key information. Slip on an Oculus Rift headset, and you're suddenly immersed in a 3D world.

What makes a great scene?

Look and feel

By intent, 3D scenes are designed to be immersive. We experience and see spaces in 3D. People viewing the content are, effectively, invited to imagine themselves within the scene as they move around. This means that the styling, or the look, of the world surrounding them can have a strong impact on how they feel about the scene in general.

For example, a city shown with dark lighting and heavy fog lends a sense of foreboding or decay, while a bright and sunny depiction of the same city, with people and cars, implies that the city is vibrant and safe—think Gotham versus Pleasantville.

Styling 3D content

The styling of the GIS content itself within the 3D scene also has a big impact on the look and feel of the scene. There are basically three choices available: fully photorealistic, fully thematic, or a combination of photorealistic and thematic.

In this web scene, the red lines represent lines of visibility that can "see" various parts of a proposed building.

Thematic scenes

Thematic views model and classify reality in a way that communicates spatial information more effectively. Thematic 3D views use common 2D cartographic techniques, such as classifications, color schemes, and relative symbol size, to simplify the real world into something that can be more readily understood. 3D scene authors create schematic, simplified representations to more effectively convey some key piece of information, particularly for scientific visualization.

For GIS users, thematic content can be an effective, and eye-catching, way to display more than just *where* something is—it can also show key *properties* about that thing. As in the example below, typhoon data points can be symbolized to show both the path of the storm and its changing wind speed.

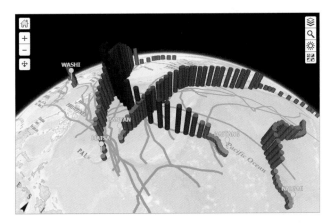

In 2005, 23 typhoons made their way through the Western Pacific Ocean region. This global-scale scene uses thematic vertical columns to describe their path and relative wind speeds, while pop-ups provide access to associated satellite photographs.

Thought leader: Nathan Shephard
The rise of the 3D cartographic scene

When people talk about seeing an amazing computer-generated 3D view, they are nearly always talking about a realistically rendered view. You know, the one with ray-tracing and ambient lighting and reflective surfaces, where it looks so much like the actual world you can almost touch it. Although this type of view is useful for conveying certain types of geographic information—such as a proposed future cityscape—it is not the right way to render everything. That is, in the same way that every map is not an aerial image, every 3D view should not be an attempt to re-create the real world.

GIS users share maps and scenes with one common goal—to communicate spatial information—and careful use of thematic symbols in 3D can be as effective, or even more effective, than similar techniques in 2D. For example, showing tree features as colored spheres on sticks (with red representing those that need to be trimmed) is much more to the point than displaying those same trees as highly realistic models covered with leaves and branches. The size of the spheres can still contain elements of the real world, such as the height and crown

Nathan Shephard is a technology evangelist and 3D GIS engineer at Esri, as well as an independent game developer.

width of each tree, but the real value of the symbols comes from their cartographic display—a simpler, more representative display that provides an immediate visual understanding of which trees are important. The advantage of using 3D is that a sphere on a stick still looks enough like a tree that you don't need to have an explicit legend saying *Tree*.

For centuries, cartographers have been limited to two dimensions. They've experimented with more effective ways of communicating spatial information through the clever use of symbols and classifications and colors. The existence of medieval, bird's-eye view maps shows that many grasped the power of the third dimension even if they didn't have the tools to fully explore it. But now suddenly, everyone has these tools, and 3D cartographers have the extra, wonderful third dimension to work with.

Video: How to author web scenes using ArcGIS Online

Who uses 3D cartography?

3D mapping and cartography have applications across a broad swath of industries and in government and academia. The examples featured here hint at the possibilities.

Take some time to click through these apps on your computer. These and many other innovative examples are collected in the ArcGIS Web Scenes gallery.

Urban planners

This 3D scene of Portland, Oregon, was created to show the impact of sunlight and visibility for a proposed high-rise development downtown.

Building and facility managers

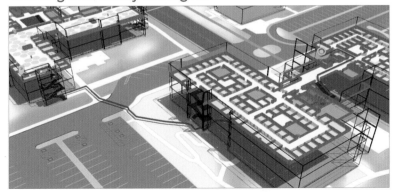

The mapping of building interiors as well as exteriors is an informative and immersive way to navigate campuses, museums, sports stadiums, and other public venues.

Social scientists

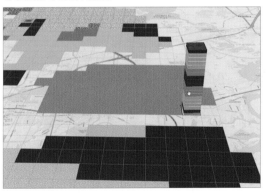

Massive datasets, such as three years' worth of crimes committed in Chicago, lend themselves to 3D visualization. In this case, the z-axis is actually used to depict time.

QuickStart

Take your maps into the third dimension with these parts of the ArcGIS platform

▸ **The ArcGIS scene viewer**
The basic ArcGIS scene viewer allows you to work immediately in 3D space. It functions with desktop web browsers that support WebGL, a web technology standard built into most modern browsers for rendering 3D graphics. Check out this gallery of scenes to verify that your browser is properly configured.

▸ **ArcGIS Earth**
This interactive globe lets you explore the world. Quickly display 3D and 2D map data, including KML, and sketch placemarks to easily understand spatial information. Download it here.

▸ **3D in ArcGIS Pro**
ArcGIS Pro is a modern 64-bit desktop application that has extensive 3D capabilities built in. You can work with 2D views and 3D scenes side by side. ArcGIS Pro is included in the free 60-day ArcGIS Trial.

▸ **Esri CityEngine**
CityEngine is an advanced tool for scenario-driven city design and developing rules for creating procedurally built cityscapes.

▸ **Terrain and basemap overlays**
Each scene starts with a basemap draped on the 3D elevation surface of the world. Zoom to your area of interest and begin to add your operational overlays.

▸ **What is the purpose of your scene?**
Before you start designing your new scene, you need to know its purpose. What is the message or information you intend to convey?

The answer to that question will help you design many elements of your scene.

- For example: Does curvature of the earth help or hinder the message (global view versus local view)?

- Will thematic styling distract from or augment the GIS information (photorealistic vs thematic layers)?

- Do users need to zoom in close to the ground (minimum surface resolution)?

- What basemap do users need draped on the ground for context (imagery, cartographic maps, thematic)?

The key point is that each of your decisions should be rooted in why you are building the scene in the first place.

▸ **2017 ArcGIS Developer Summit Video**

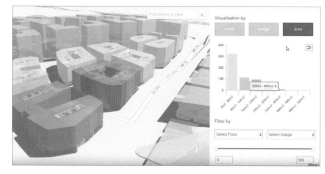

Quick tour of new products and capabilities in 3D

Learn ArcGIS lesson

Create a 3D Web Scene of Underwater Beach Restoration Efforts off the Coast of Palm Beach County, Florida

The beaches and inlets along the coast of Palm Beach County, Florida, contain a delicate ecosystem teeming with flora and fauna. However, beaches are unstable by nature. Beach sand is washed away by ebbing tides and occasional storms. Coastal areas require frequent restoration and maintenance. Sand is excavated, or dredged, from shallow areas or inlets to replenish eroded beaches, while artificial reefs are constructed to protect the shoreline. To manage these complex restoration efforts, proper monitoring and mapping is essential.

Overview

In these lessons, you'll help the Palm Beach County beach restoration efforts by mapping some of the county's major beaches and inlets as part of a presentation for both the public and policy makers. To emphasize bathymetric features and topography, you'll create your map in 3D using the ArcGIS scene viewer. Begin by adding layers depicting reefs, sediment, and dredging areas to a new scene. Then capture slides of key areas so that users can quickly navigate to the locations you want to emphasize. Finally, create a web app to share with others.

▸ **Build skills in these areas:**
- Navigating a scene
- Adding layers to a scene
- Making layer groups to organize data
- Capturing slides
- Creating a 3D web app

▸ **What you need:**
- Publisher or administrator role in an ArcGIS organization
- Estimated time: 20 to 40 minutes

Start Lesson

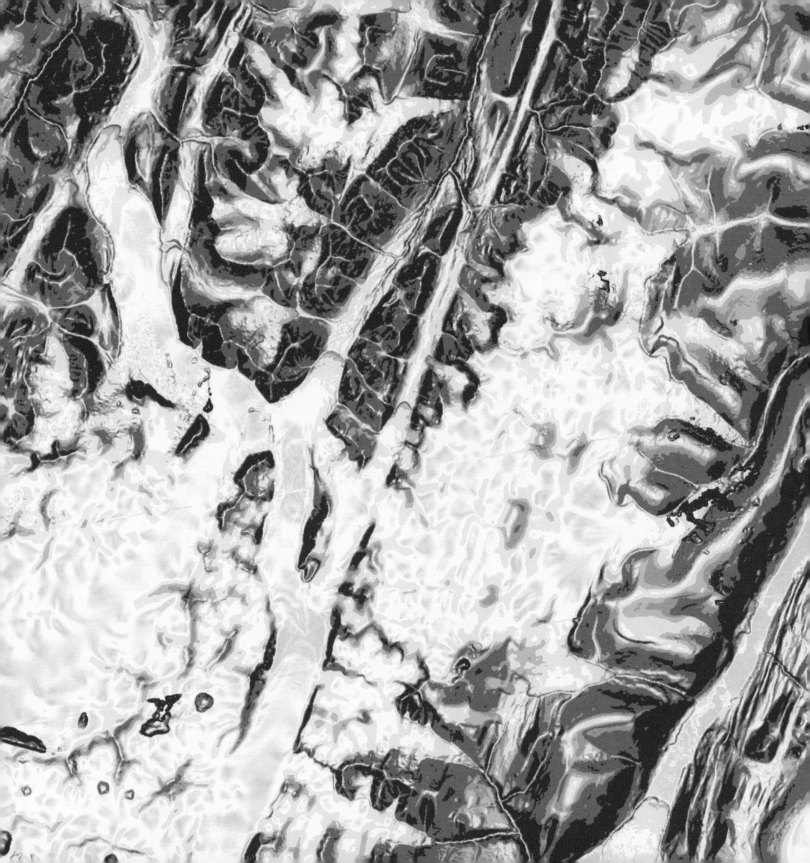

Murray Ice Cap, Ellesmere Island

The Power of Apps
Focused tools that get the work done

With billions of smartphones, tablets, laptop computers, and other Internet-connected devices in use worldwide, stand-alone apps have captured the world's attention. GIS apps in particular have transformed how people think about geography. Every map has an interface—a user experience that brings that particular map into use. These experiences bring GIS to life for all kinds of uses, including the Arctic Elevation Explorer app shown here featuring the latest high-definition terrain measurements.

The rise of spatially intelligent apps

Apps are lightweight computer programs designed to run on the web, smartphones, tablets, and other mobile devices. And GIS apps are a special breed; they're map-centric and spatially aware.

Today apps are ubiquitous. Billions of people worldwide run them in their web browsers, on computers, and on their mobile devices. Creating interesting, geographically aware apps is now easily within your reach. From the intuitive story map apps and Web AppBuilder for ArcGIS® to the app collection for your smartphone and tablet, the technology required to deploy highly effective apps that can extend the reach of GIS to new audiences is transforming the power and reach of your GIS throughout the world.

Apps are often built around targeted workflows that deliver streamlined user experiences. They're designed to guide users through specific stories or tasks, show just the focused information that is required for that task, and enable the efficient communication of your message.

This chapter shows where apps come from and how you can start to create your own. It explores some of the innovative ways that apps are being used to do real work with ArcGIS. You'll discover ArcGIS apps for the work you do, no matter what the task or the device. Need to collect data in the field? There's an app. Need to share your data with the public? There's an app for that, too. Whether you're managing a mobile workforce, creating a geolocation start-up, or looking for innovative ways to share your information in a useful and meaningful way, apps are your pathway to achieving that.

Mt. Everest

29,029 feet | 8,848 meters

The highest point on Earth. First climbed by Edmund Hillary and Tenzing Norgay in 1953, Everest's summit has been reached by nearly 4,500 people and sometimes experiences traffic jams. Yet it remains a potentially deadly mountain: on average five climbers die on Everest each year.

ArcGIS everywhere
Apps take ArcGIS where you go and where your users go

With mobile phones and devices, your GIS maps and apps go with you wherever you go. That's a big idea. Your phone is a sensor with continuous location awareness coupled with the capacity to take geotagged photos and collect data by location. The integration of the smartphone and GIS carries many implications.

Use GIS in the field
You can use your smartphone to capture geotagged photos and videos in the field, and then use them to tell and share your stories. You can collect data in the field and update your enterprise information.

Connect to your enterprise GIS
Your phone can also be used to access enterprise information for your location so that you have deeper knowledge and awareness. You can use maps to navigate and assist you in the field and to perform a variety of work tasks—collect information or perform a survey, and then sync your results with your GIS in the office.

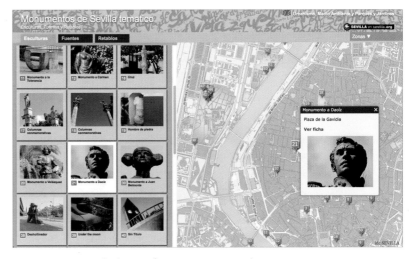

Use geotagged photos from your smartphone to create a story map like this example of the Monuments and Art of Seville, Spain, or if you're in Seville, load the app on your smartphone to help you find the monuments.

This app features live CCTV data showing storm drain conditions in Naperville, Illinois. Having the information packaged in a way that is consumable by a web browser or mobile devices put it in the hands of people who need it to do their work.

Solve a problem with an app

Every online GIS map has an interface—an experience that enables people everywhere to apply GIS. Interestingly, these experiences are actually apps—much like the ones you use on your smartphones every day. The following pages describe almost a dozen different things that can be done using GIS apps. As you read through them, think about what it is that you need to do that could fit into one of these patterns. And then consider the possibilities.

Tell a story

You can author a story map (the focus of chapter 3) fairly easily by choosing from the many Esri Story Maps narrative styles offered in ArcGIS to engage and inspire your audience. Story map apps combine maps with rich narratives and multimedia content that connect with an audience and keep them engaged.

This story map tells the history of the City of Greenville, South Carolina's, master plan for revitalizing downtown.

Engage with people

GIS apps offer an exciting and engaging way to publish your geographic information. When you can interact with a live map, communicate back to the GIS of the world, and have the app follow you and alert you when you get certain places, all of a sudden you've got something powerful and engaging because the experience is personal and familiar to you.

iGeology (available on Google and Apple app stores) is a free smartphone app that lets you take over 500 geological maps of Britain wherever you go to discover the landscape beneath your feet.

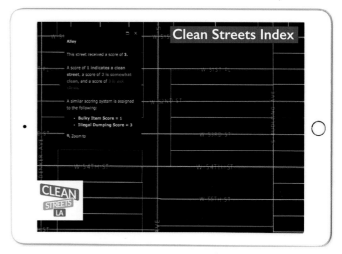

The Los Angeles Clean Streets Index, the first of its kind, is a grading system for every street in the city. It is a call to action aimed at connecting citizens with the city's Bureau of Sanitation to improve the health and cleanliness of their neighborhoods.

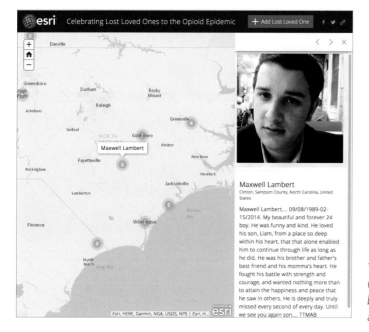

This crowdsource map based on the Esri Story Map Crowdsource app lays bare the sorrow and pain caused by the dramatic rise in deaths tied to prescription drugs and heroin use in the United States.

Take GIS into the field

Anywhere you see organizations and people doing work in the field, there's a potential for applying GIS to improve coordination of operations and achieve greater efficiencies and cost savings. Here are some examples of how ArcGIS is being used in the field.

Environmental monitoring

Pronatura Noroeste, the largest nonprofit conservation group in Mexico, collects marine field observation data by means of a GIS app, replacing the error-prone use of paper forms in remote marine environments. Nontechnical field inspectors were able to easily configure the smart Survey123 app for ArcGIS with no programming skills. The goals of increased data consistency, efficiency, and accuracy were achieved.

Archaeology

At one of Öland's ancient ring forts, archaeologists from the Kalmar County Museum of Sweden have used ArcGIS and field GIS tools to make several discoveries and observations about this fort originating from the ancient Roman Empire. This has provided an intriguing view into the past back to the fifth century. You can follow them by video as they explore their site using ArcGIS.

Collect high-resolution, up-to-the-minute aerial photography

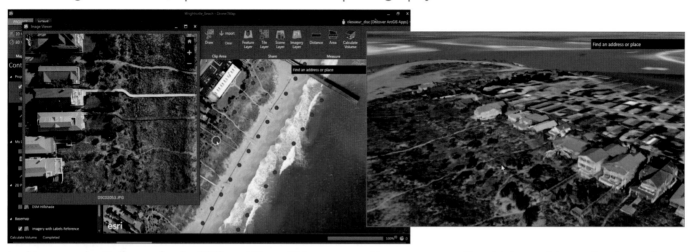

In this example, Collector for ArcGIS® and Drone2Map are used in concert to collect high-precision orthophotos for a beach area in Wilmington, North Carolina. Collector was used to capture a network of survey control points with 3 cm accuracy. This network of ground control points was used for georeferencing high-resolution photography and elevation collected from a low-altitude drone mission along the beach.

Resources to learn more about GIS field apps

Navigator for ArcGIS®

Workforce for ArcGIS®

Collector

Survey123 for ArcGIS®

Explorer® for ArcGIS

Drone2Map

Monitor and manage your operations

In today's world, GIS people think of GIS as the way to create, monitor, and manage their "digital twins," meaning the computerized version of their everyday world. Using GIS as a geospatial framework for tracking and monitoring your work makes a lot of sense. Digital twins use data from sensors installed on various assets and resources to continuously monitor their status, working condition, and location. Geography provides a universal way to organize and manage operations using these map locations and sensor feeds from the field.

Most organizations stay on top of their operations by monitoring, tracking, and reporting real-time data feeds. For example, many commercial companies track their sales and their competition; epidemiologists track disease; first responders monitor and manage events and incidents; wildlife scientists track animals; farmers sense the state of their crops; and meteorologists monitor and forecast the weather. Meanwhile, all kinds of organizations track and manage their mobile workforces.

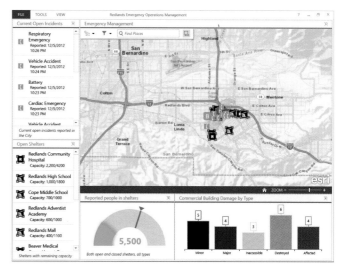

Organizations use the Operations Dashboard for ArcGIS® to monitor deliveries, services, people, vehicles, and other assets, anywhere in the world.

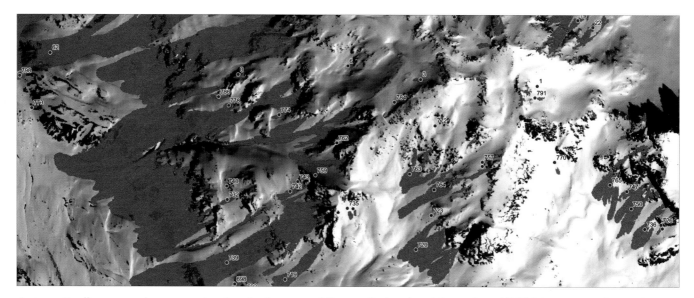

Automatically mapped snow avalanches in the area of Davos, Switzerland, based on ADS80 images at the end of the winter of 2012–2013. These automated mapping alert systems are integrated into the region's emergency response system.

Answer questions using spatial analysis

Geographic insight is often the best way to answer pressing questions. Overlaying multiple layers on a map and analyzing them with advanced spatial models can highlight relationships that are not otherwise apparent. Knowledge workers using ArcGIS have the power to build models to address almost any question, from where to locate a new facility to finding areas most at risk, and deploy them as apps.

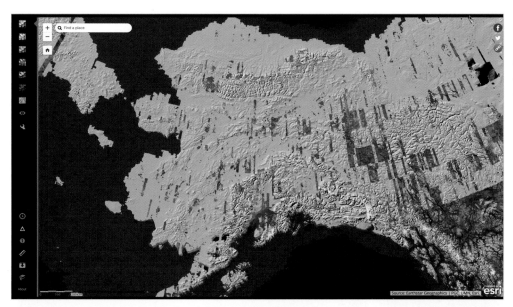

New models have been added to the ArcticDEM project maps, a public-private initiative to produce high-resolution, high-quality digital elevation models (DEMs) of the Arctic.

The USDA Forest Service monitors and forecasts the distribution and extent of forest stands for various timber species. It uses this data to understand the science and issues confronting the best use of timberlands in a world whose climate is changing.

Esri publishes a tapestry dataset of lifestyle information for the United States. Type a ZIP Code and explore the demographics for your area of interest.

Apps expand the reach of your GIS
Extending the reach of your GIS with apps

Outdoor recreation

Donegal, Ireland, is a hill walkers' haven at any time of the year, full of high peaks, panoramic vistas, quiet tracks, and clean air. This 3D web scene depicts highlighted tours and serves as a gateway to the Donegal Map Portal, an extensive collection of GIS data and map resources.

GIS inside Excel

Maps for Office brings the mapping component to Microsoft Excel.

Planning

GeoPlanner for ArcGIS is a simple-to-use app that applies sophisticated spatial analysis to support planning scenario comparisons and workflows.

Thought leader: Jeff Shaner
On the scene at the Deepwater Horizon oil spill

During the 2010 Deepwater Horizon oil spill in the Gulf of Mexico, I was part of a team sent by Esri to assist our customers among the several different emergency response agencies that were operating at the scene. The situation was somewhat intense, and we were in meetings where a lot of information was flying around—not all of it accurate or timely. Dozens of teams were in the field—to monitor the developing situation, collect data, and conduct environmental surveys. The data collection effort was still largely paper-based, and coordination among all the teams was difficult.

The problem wasn't a lack of maps or GIS. These agencies were already among our most sophisticated users. The problem was in appropriately sharing and keeping each other updated as new information flowed into the operations center. I witnessed how the teams gathering data out on the water and along the shoreline were struggling early on to collect their information and make it accessible so that it could be acted upon.

Within a week, thanks to a lot of hard work by scores of technologists, emergency response professionals, and staff from British Petroleum, many of the pieces were falling into place, and we saw GIS beginning to be used for mobile data collection and communication. These teams began to share maps, data, videos, and photos, enabling responders to better coordinate with emergency command centers and ensure high

Jeff Shaner is a Product Manager at Esri focused on the development of new mobile and web technology offerings.

levels of situational awareness. As tragic as the event was, I did see how much more effective these teams were by using mobile GIS.

When we returned to Esri, we used this firsthand experience to help guide our product development forward, by applying many of the ideas that came out of those frantic weeks. One idea borne from this initiative became the first generation of our Collector for ArcGIS app.

It's heartening for me to know that these same agencies are now equipped with Collector for ArcGIS, among a whole suite of new apps that equip response teams with more efficient rescue and recovery capabilities.

Where do apps come from?

You can get usable apps for your own devices and those of your audience from a number of sources. They range from the off-the-shelf apps on popular app stores from Esri and other developers in the business community—and "roll your own" apps that you configure and publish using templates and builders—to fully customized solutions built by developers with software development kits (SDKs) and application program interfaces (APIs).

ArcGIS apps

ArcGIS includes a suite of apps that are ready to go and free to use if you have an ArcGIS℠ Online Organizational Account. Mapping apps, such as Explorer for ArcGIS (on the Apple iOS platform), provide a way to manage a collection of data.

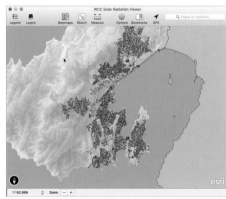

ArcGIS Marketplace

At the ArcGIS Marketplace you can obtain apps and data services not only from Esri but also its distributors and partners. All the apps in the Marketplace are built to work with ArcGIS Online, so they can easily be shared with ArcGIS Online groups and users within your organization.

Explorer for ArcGIS

Industry-specific apps

ArcGIS Solution apps support a range of industries such as local and state government, emergency management, utilities, telecommunication, military, and intelligence. You can utilize this rich collection of preconfigured apps to jump-start your enterprise implementations of ArcGIS.

Create your own apps

ArcGIS provides developer tools for app builders. Coding your own apps requires more effort, but provides the most flexibility. The two studio-based application tools below provide a developer's workbench that enables you to compose your own apps and minimize the need to write a lot of your own code.

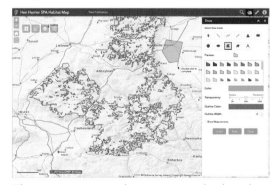

This Hen Harrier Habitat app was built with Web App Builder for ArcGIS.

FAA used AppStudio for ArcGIS to build its open data site.

Case study: US Geological Survey

In 2009, the US Geological Survey (USGS) began the release of a new generation of topographic maps (US Topo) in electronic form and, in 2011, complemented them with the release of high-resolution scans of historical topographic maps of the United States dating back to 1882. The topographic map remains an indispensable tool for everyday use in government, science, industry, land management planning, and recreation.

Historical maps are snapshots of the nation's physical and cultural features at a particular time. Maps of a particular area can show how the area looked before development and provide a detailed understanding of change over time. Historical maps are often useful to scientists, historians, environmentalists, genealogists, citizens, and others researching a particular geographic location or area. The USGS, with help

from Esri, has created an app that lets you view this extensive collection of topographic maps in one central location. The USGS Historical Topographic Map Explorer makes it easier to dig into and enjoy this library of more than 178,000 historical maps in a web app that organizes the maps by space, time, and map scale.

Using the USGS Historical Topographic Map Explorer is easy—just follow the numbered steps at the left of the interface pane. The choices you make will update the adjacent map view.

Here's how to get the most out of the app:

- Find the area you want to explore
- Use the timeline to select the maps
- Compare the maps

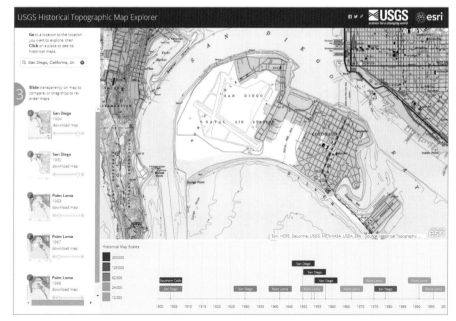

The USGS was created in 1879, charged with the "classification of the public lands, and examination of the geological structure, mineral resources, and products of the national domain." This simple-to-use app compiles this impressive body of historical mapping work into an engaging user experience.

QuickStart

Use out-of-the-box apps, build apps without having to write any code, or code your own apps from scratch

▸ **Use Esri's ArcGIS apps**
These out-of-the-box apps are ready for you to use for your needs:

Explore maps—Explorer for ArcGIS: take a tour.
Collect data—Collector for ArcGIS: try collecting a damage assessment.
Manage operations—Operations Dashboard for ArcGIS: monitor a city's emergency response to an earthquake.
Analyze demographic data—Esri® Community Analyst: see an overview video of the app.
Analyze and evaluate scenarios—GeoPlanner for ArcGIS: get more information.
Integrate with your business data—Location Analytics: find the perfect app for your business.

▸ **Find apps in the ArcGIS Marketplace**
Esri has created a marketplace for you to find apps developed by Esri, business partners, and others, all built on top of ArcGIS.

▸ **Developers.arcgis.com**
If you know the difference between API and SDK, head to the ArcGIS for Developers website or GitHub.

▸ **Build your own app**
If the out-of-the-box apps don't do what you need, why not build one yourself?

No coding required—Web AppBuilder for ArcGIS: make your first app in five minutes.
Configure template apps—ArcGIS Solutions: explore solution templates to jump-start your projects.
Code a web app—ArcGIS® API for Javascript™: use the API to build your first web app.
Code a native app—ArcGIS® Runtime SDKs: see the power of native apps with ArcGIS Runtime developer kits.

Share your apps

Web Apps
1.   Select the app.
2.   Configure it.
3.   Save it and share it in ArcGIS Online or Server.

Native Apps
1.   Find the app you want to share in the iTunes App Store or Google Play.
2.   Share the URL with your audience.

Learn ArcGIS lesson

Get Started with Workforce, Collector, and Navigator for Hydrant Inspections

The City of San Diego, like all big cities, relies on fire hydrants to help put out fires, keeping people and property safe. To ensure that the hydrants work when they are needed, the San Diego Fire Department regularly inspects them, but until recently keeping track of this work has been tedious, paper-based activity.

In this exercise, you'll help improve the system of inspections using GIS technology. Acting as a few different members of the City of San Diego and the San Diego Fire Department, you'll create and manage hydrant inspection assignments as well as completing that work in the field. Your goal is to make sure the fire hydrants around the convention center are all inspected before the end of summer.

▶ Overview
Your fire captain has assigned you some hydrant inspections, so you must make sure the hydrants will work when you need them. Your fire captain is using a Workforce project to manage your work, and you'll use the mobile app to view and complete your assignments. Along the way, you'll use Navigator to route to the hydrants you must inspect.

▶ Build skills in these areas:
- Creating a Workforce project
- Creating and managing assignments in Workforce
- Using Workforce and Navigator to complete assignments

▶ What you need:
- Publisher or Administrator role in an ArcGIS organization
- Workforce for ArcGIS and Navigator for ArcGIS installed on a mobile device
- An ArcGIS organization account licensed for Navigator for ArcGIS
- Estimated time: one to 1½ hours

Start Lesson

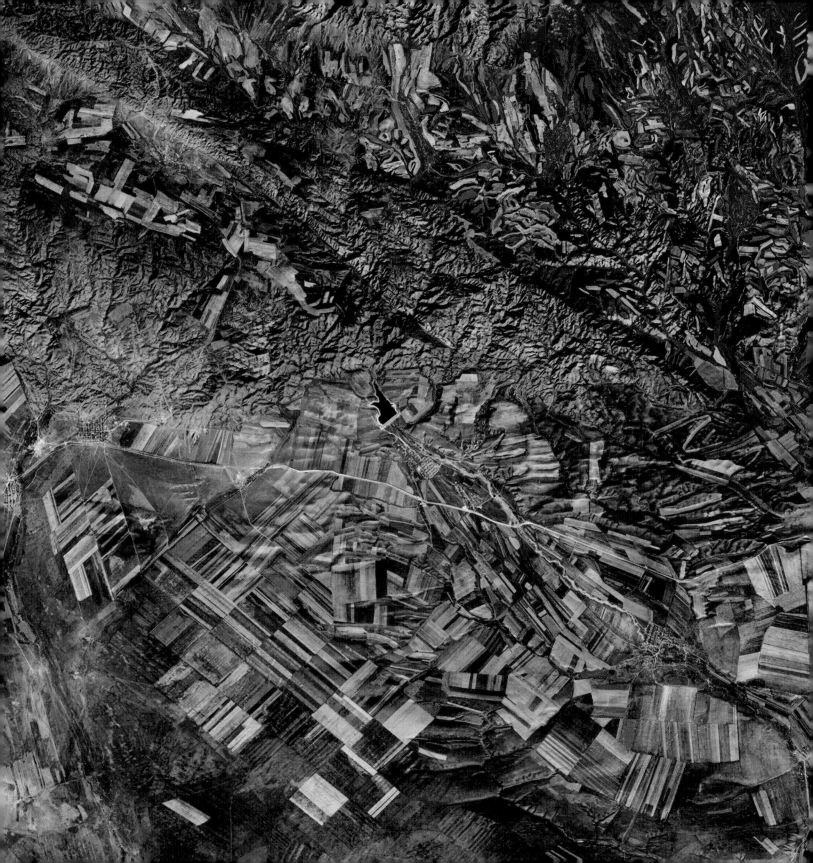

08

Imagery Is Visible Intelligence
A geographic Rosetta stone

GIS is both intuitive and cognitive. It combines powerful visualization and mapping with strong analytic and modeling tools. Remotely sensed earth observation—generally referred to in GIS circles simply as *imagery*—is the definitive visual reference at the heart of GIS. It provides the key, the geographic Rosetta stone, that unlocks the mysteries of how the planet operates and brings them to life. When we see photos of Earth taken from above, we understand immediately what GIS is all about.

Imagery deepens understanding
Seeing is not only believing, it is also perceiving

The story of imagery as an Earth observation tool begins with photography, and in the early part of the 20th century, photography underwent extraordinary changes and social adoption. Photos not only offered humanity a new, accessible kind of visual representation—they also offered a change in perspective. The use of color photography grew. Motion pictures and television evolved into what we know today. And humans took to the sky flying in airplanes, which, for the first time, enabled us to take pictures of the earth from above. It was a time of transformation in mapping and observation, providing an entirely new way of seeing the world.

World War II: Reconnaissance and intelligence gathering

During World War II, major advances in the use of imagery for intelligence were developed. The Allied Forces began to use offset photographs of the same area of interest, combining them to generate stereo photo pairs for enhancing their intelligence gathering activities. In one of many intelligence exercises called Operation Crossbow, pilots flying in planes—modified so heavily for photo gathering that there was no room for weapons—captured thousands of photographs over enemy-held territory. These resulting collections required interpretation and analysis of hundreds of thousands of stereo-pairs by intelligence analysts.

These 3D aerial photographs enabled analysts to identify precise locations of highly camouflaged rocket technology developed by Germany. This was key in compromising the rocket systems that were targeting Great Britain, thus saving thousands of lives and contributing to ending World War II. The BBC did a comprehensive documentary on this subject (*Operation Crossbow: How 3D Glasses Helped Defeat Hitler*).

Stereoscopic imagery was instrumental in identifying the facilities of Nazi rocket programs. The photo above shows stereo glasses used for viewing offset photo pairs. This June 1943 photograph (left) was the first to reveal functional weapons. Two V2 rockets 40 feet long are seen lying horizontally at (B), but only in December was it realized that the structure at (C) was a prototype flying-bomb catapult.

1969: Dawn of extraterrestrial man
The first humans explore our moon

In the early 1960s, the majority of people would probably have said it was impossible for a human being to walk on the moon. But in July 1969, televised images transmitted to Earth from the moon showed Neil Armstrong and Buzz Aldrin bounding across the lunar surface, proving that moon walking was more than conceptually possible—it was happening right before our eyes. Seeing was believing.

When Armstrong, Aldrin, and the ensuing lunar astronauts pointed their cameras back at Earth, an unexpected benefit became apparent: humanity now had a completely new perspective about our home planet—heralding the adoption and use of Earth imagery (see "Earth from Space" on page 129).

In December 1972, Apollo 17 astronauts captured this iconic photo of Earth from space—the famous "Blue Marble" photograph, offering humanity a new perspective about Planet Earth and our place in the universe.

Astronaut Buzz Aldrin, from the Apollo 11 mission, on the moon in July 1969. Photo by Astronaut Neil Armstrong (visible in Aldrin's face shield).

1972: The Landsat program
Providing the first satellite images covering Earth

In 1972, the same space technology that was developed to put humans on the moon led to the launch of the first Landsat satellite. The Landsat mission gave us extraordinary new kinds of views of our own planet. This was a breakthrough system and the first civilian-oriented, widely available satellite imagery that not only showed us what was visible on Earth—it also provided a view of invisible information, unlocking access to electromagnetic reflections of our world. We could see Earth in a whole new way.

This persistent earth observation program continues to this day along with hundreds of other satellites and remote sensing missions as well. Nation states and, more recently, private companies have also launched numerous missions to capture earth imagery, allowing us to continuously observe and monitor our planet.

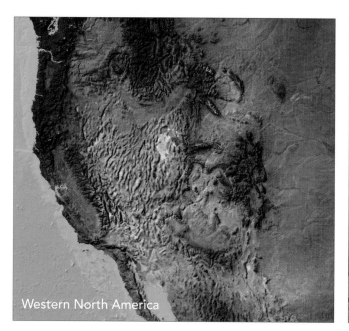
Western North America

New Orleans

Landsat sensors have been continually generating and sharing pictures of the earth since the 1970s. Early on, scientists were excited by the valuable new perspectives being generated. Today, huge numbers of satellites image the earth thousands of times daily, creating a massive and virtual image catalog of our planet. Web GIS is tapping into these images to enable practitioners to address a broad array of questions and challenges that we face as Earth's stewards.

Show me my home!
2005: The human era of GIS begins

Little more than a decade ago, seemingly the whole world snapped awake to the power of imagery of the earth from above. We began by exploring a continuous, multiscale image map of the world provided online by Google, Microsoft, and other companies. A combination of satellite and aerial photography, these pictures of Earth helped us experience the power of imagery, and people everywhere began to understand some of what GIS practitioners already knew. We immediately zoomed in on our neighborhoods and saw locational contexts for where we reside in the world. This emerging capability allowed us to see our local communities and neighborhoods through a marvelous new microscope. Eventually, naturally, we focused beyond that first local exploration to see anywhere in the world. What resulted was a whole new way to experience and think about our planet.

These simple pictures captured people's imagination, providing whole new perspectives, and inspired new possibilities. Today, virtually anyone with Internet access can zero in on their own neighborhood to see their day-to-day world in entirely new ways. In addition, people everywhere truly appreciate the power of combining all kinds of map layers with imagery for a richer, more significant understanding.

Almost overnight, everyone with access to a computer became a GIS user.

Initially, we zoomed in on our homes and explored our neighborhoods through this new lens. This experience transformed how people everywhere began to more fully understand their place in the world. We immediately visited other places that we knew about. Today, we continue by traveling to faraway places we want to visit. Aerial photos provide a new context from the sky and have forever changed our human perspective. This "Show me my home" app provides a simple user experience for viewing global imagery at any scale..

Imagery expands your perspective
Seeing the visible, the invisible, the past, and into the future

Seeing is believing. Observing the world in colorful imagery is informative and immediate, delivering stark visual evidence and new insights. Imagery goes far beyond what our own eyes are capable of showing us—it also enables us to see our world in its present state. And it provides a means to look into the past as well as forecast the future, to perceive and understand Earth, its processes, and the effects and timelines of human activity. Amazingly, imagery even allows us to glimpse the invisible, to see visual representations of reflected energy across the entire electromagnetic spectrum, and to thus make more fully informed decisions about the critical issues facing Earth and all its life-forms.

Understanding patterns
Global imagery is collected continuously, enabling us to witness our world in action. By combining images from across the span of time, we can begin to visualize, animate, analyze, and understand Earth's cycles, where we came from, and where we are going.

Weather provides "breathing ranges of snow and ice" for the planet, delivering precious water that enables and sustains all living things. This image shows the seasonal weather cycles of precipitation across the North American continent.

Seeing beyond the visible
Imagery enables us to see beyond what our human eyes perceive, providing new scientific perspectives about Earth. Satellites have sensors that measure nonvisible information, such as infrared energy, across the electromagnetic energy spectrum that enables us to generate and analyze a multitude of new terrestrial views of our world.

This false color image over North Africa shows dry and wet areas of vegetation moisture by looking at the near infrared (band 5) and shortwave infrared (band 6). Warmer colors reflect arid areas in the map. The striping pattern shows Landsat 8's scene footprints, illustrating its continuous orbit of Earth and revisiting every scene location roughly every 16 days.

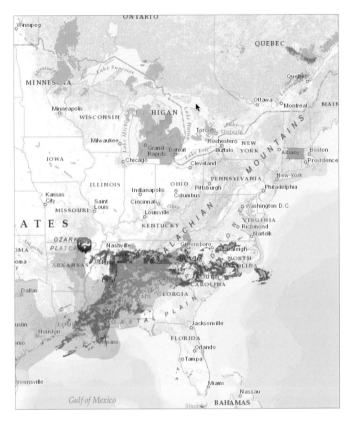

Forecasting and reporting weather

Advances in imaging and weather observations over the last decade have resulted in a dramatic increase in the accuracy and precision of meteorologic forecasts. GIS integration of climate data for operations management has expanded to benefit farmers, emergency response teams, school districts, utilities, and many others. The sensors range from global weather satellites to ground-based local instruments that allow experts to monitor and forecast meteorological events like never before. The sensor network has become hyperlocal, allowing continuous forecasting of weather events in our communities. We can now access an accurate forecast for our neighborhood for the upcoming hour.

This radar-derived layer for the US mainland from AccuWeather shows precipitation in near real time. These near real-time observations, along with weather forecasting, are managed using image observations.

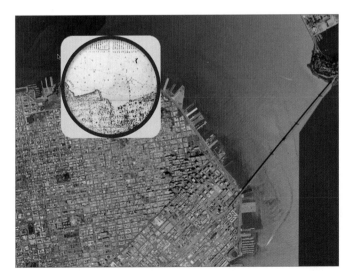

Beyond the apparent

Imagery lets us to peer into the past as well as combine historic views with current imagery. Imagery comes in a simple format, allowing it to be easily overlaid with other maps and images into a kind of layered "virtual sandwich."

This unique spyglass app shows how the city of San Francisco expanded beyond the historic coastline settlement. With the city's location along the San Andreas Fault, expanding into the bay had its unique challenges requiring construction engineers to push pilings down about 200 feet to bedrock.

Thought leader: Lawrie Jordan
ArcGIS now includes a complete image processing system

A special relationship has always existed between GIS and remote sensing, and it goes back to the very beginning of our modern information technology. In the 1960s and 1970s, computer systems for GIS were big, expensive, and very slow mainframes using punched cards, but nearly all the foundation data layers in these early systems came either directly or indirectly from imagery. Right from the start, GIS and remote sensing were complementary, like two sides of the same coin. They coevolved together.

In 1972, a revolution happened with the launch of Landsat—the first commercial Earth observation imaging satellite. It continuously orbited the earth and captured a new image of the same spot about every 16 days. Because it was so high up, it gave us an entirely different picture of our planet and its patterns. It provided not only a new view; it gave us a new vision of the possibility of what GIS could become. And it started a revolution in commercial Earth observation that continues today and is exploding now with hundreds—and soon thousands—of smaller satellites, microsatellites, video cameras from space, high-altitude drones, and more.

So where are GIS and remote sensing—these two close allies for more than 50 years—going next?

For one thing, there's a big emphasis now on simplicity and speed. It's clear that the future belongs to the simple and quick. We're seeing that modern technology is harnessing this amazing array of globally distributed sensors into what is popularly referred to as the Internet of Things, a vast collection of dynamic, live information streams that are feeding into and becoming the heart of Web GIS. Plus, this network operates in real time, giving us access to what we might call the "Internet of All My Things"—and all on our own devices through a new geoinformation model.

Although the technology powering this concept is advanced, we comprehend it in practice because we understand pictures. Einstein famously said, "If I can't see it, I can't understand it." We all know something when we can see it.

And now, all these rapidly changing developments combining imagery and spatial analyses are opening up new chapters in the history of GIS, as society is awakening to the power of geography and the intuitive understanding that imagery helps us "see" in all its forms.

We like to say that the map of the future is an intelligent image.

Lawrie Jordan is Esri's Director of Imagery and Remote Sensing. He is a pioneer in the field of image processing and remote sensing.

 Video: The map of the future is an intelligent image

Imagery has so many uses
A range of applications

By now, it's apparent that imagery enables whole new perspectives and insights into your world and the issues you want to address. Imagery also has numerous advantages and capabilities.

Almost daily access to new information
Image collection is rapid and increasing. And access to imagery is increasingly becoming more responsive. Many satellites and sensors are already deployed with more coming all the time, collecting new data, adding to a continuous collection effort—a time series of observations about our planet. These image collections are enabling us to map, measure, and monitor virtually everything on or near the earth's surface. All of us can quite rapidly gather much of the data that we need for our work. Imagery has become our primary method for exploration when we "travel" to other planets and beyond. We send probes into space and receive returns primarily in the form of imagery that provides a continuous time series of information observations. And it enables us to derive new information in many interesting ways.

Looking back in time
The use of aerial imagery is still relatively young. Although imagery began to be used only in the 20th century, it is easy to compare observations for existing points in time that reside in our imagery collections. In addition, we can overlay imagery with historical maps, enabling us to compare the past with the present.

Data collections are becoming richer every day
Imagery is leading to an explosion of discovery. Many imagery initiatives are growing, expanding, and adding to image databases for our areas of interest. ArcGIS includes a complete image processing system, enabling the management of increasingly large, dynamically growing Earth observations. This points to the immediacy of imagery and its capacity for easy integration, enabling all kinds of new applications and opportunities for use—things like before-and-after views for disaster response, rapid exploration of newly collected imagery, image interpretation and classification, and the ability to derive intelligence. Over time, many of these techniques will grow in interesting new ways, enabling deeper learning about our communities, the problems and issues we face, and how we can use GIS to address them.

Powerful analytic capabilities
Imagery and its general raster format enable rich analysis using ArcGIS. And, in turn, these analyses enable more meaningful insights and perspectives about the problems you want to address.

Working together at last
Combining GIS and image processing provides synergy

Imagery in all its variations uses one of the key common data formats in GIS: *rasters*. Rasters are one of the most versatile GIS data formats. Virtually any data layer can be conveyed as a raster. Using rasters, you can combine all kinds of data with your imagery, enabling integration and analytics.

Rasters provide a host of useful GIS data layers

Rasters, like any digital photo, provide a data model that covers a mapped area with a series of pixels or cells of equal size that are arranged into a series of rows and columns. Rasters can be used to represent pictures as collections of pixels, surfaces such as elevation or proximity to selected features, all kinds of features themselves (in other words, points, lines, and areas), and time series information with many states for each time period.

Classified land cover and land use

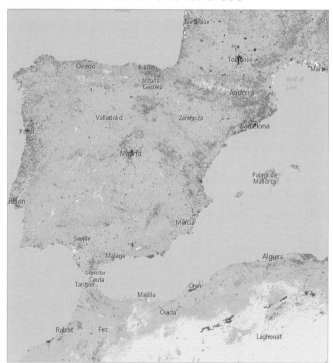

Land cover around the western Mediterranean, from a global raster dataset from MDA Information Systems of the predominant land characteristics at 30-meter resolution.

Distance to water

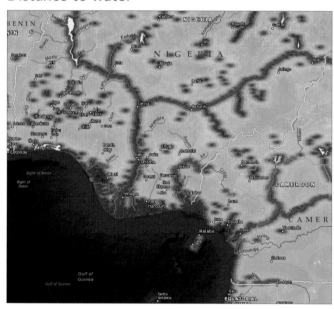

A proximity map showing distance from each cell or pixel to a reliable water source in a portion of West Africa. Water access is vital for humans as well as wildlife habitat. Streams are overlaid on the distance grid. Cells in the grid that are closest to water are darker blue. Colors change as the distance from water increases.

Three-dimensional scenes

Mont Blanc or Monte Bianco in the Alps between France and Italy is included in this app (linked above), which features a 3D tour of interesting sites from around the world.

Elevation expressed as shaded relief

Global elevation displayed as shaded relief. This is part of a global elevation layer compiled from the best available sources worldwide.

Oblique imagery photos

Oblique imagery provides a special perspective view of real-world features, presenting natural detail in 3D and enabling interpretation and reconnaissance.

Time series information

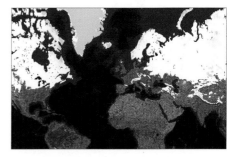

A snapshot of a time-enabled image map of monthly snowpack observations from the NASA Global Land Data Assimilation System (GLDAS). This map contains cumulative snowpack depths for each month from 2000 through 2015.

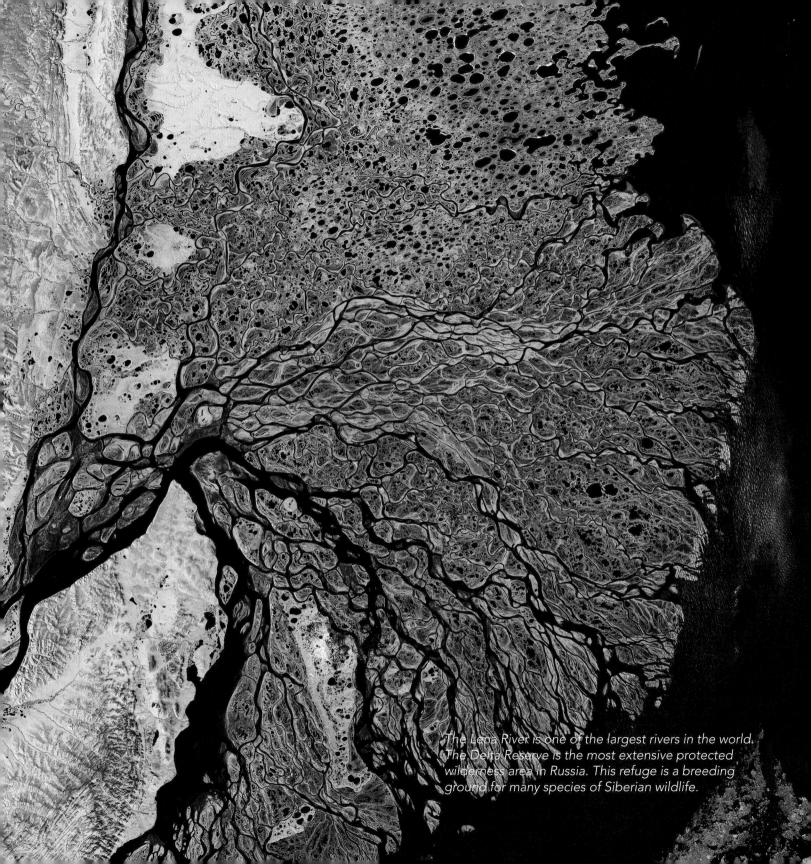

The Lena River is one of the largest rivers in the world. The Delta Reserve is the most extensive protected wilderness area in Russia. This refuge is a breeding ground for many species of Siberian wildlife.

Imagery is beautiful
Both informative and sublime

While imagery provides whole new perspectives, profoundly shaping our understanding, it's also clear that imagery provides exquisite views of our world—truly stunning and beautiful works of art. They astonish and amaze us, tapping into our emotions and the wonder of our world and new worlds we seek to discover and explore. It's no accident that the US Geological Survey maintains a collection of *Earth as Art*.

Along Greenland's western coast, a small field of glaciers surrounds Baffin Bay.

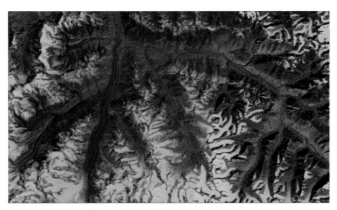

Soaring, snow-capped peaks and ridges of the eastern Himalayas create an irregular red-on-blue patchwork between major rivers in southwestern China.

Snow-capped Colima Volcano, the most active volcano in Mexico, rises abruptly from the surrounding landscape in the state of Jalisco.

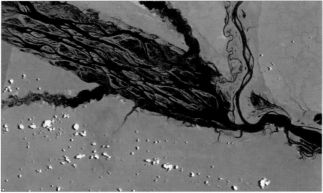

Fed by multiple waterways, Brazil's Negro River is the Amazon River's largest tributary. The mosaic of partially submerged islands visible in the channel disappears when rainy season downpours raise the water level.

Earth from space
The power of a single image

In the run-up to the Apollo moon landings, Apollo 8 was the first mission to put humans into lunar orbit. And on Christmas Eve 1968, coming around from the far side of the moon during their fourth orbit, Apollo 8 commander Frank Borman exclaimed, "Oh my God, look at that picture over there! Here's the earth coming up! Wow, that is pretty!" Fellow astronaut Bill Anders grabbed his Hasselblad camera and shot this now-famous image of Earth rising above the moon.

In his book *Earthrise: How Man First Saw the Earth*, historian Robert Poole suggests that this single image marked the beginning of the environmental movement, saying that "it is possible to see that *Earthrise* marked the tipping point, the moment when the sense of the space age flipped from what it meant for space to what it means for Earth." The power of imagery can be neatly summed up in the story of this single photograph. Images can help us better understand our planet, drive change, create connections—and in some cases even start a movement.

It's one of the most frequently reproduced and instantly recognizable photographs in history. The US Postal Service used the image on a stamp. Time magazine featured it on the cover. It was—and still is—"the most influential environmental photograph ever taken," according to acclaimed nature photographer Galen Rowell.

Mapping the solar system
An effort reflecting humanity's seeking spirit

Since the first moon shots, astronaut-photographers from the world's space agencies have also been turning their lenses *away* from Earth. GIS people, being the science fanatics they often are, have of course found ways to map planetary bodies other than our home planet. In 2015, NASA announced to the world that multispectral imagery taken from Mars-orbiting sensors had definitively ascertained the one-time presence of moving water on Mars—a milestone not lost on the GIS and image analysis community.

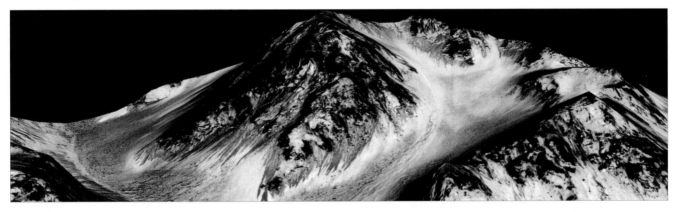

3D visualization of the hyperspectral imagery data that changed our perception of the planet Mars.

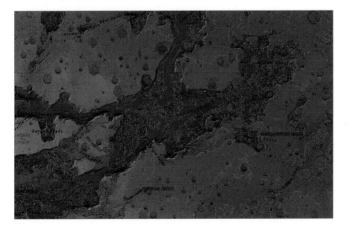

This map paints a picture of the dramatic geography of Mars and all the surface missions that humans have carried out in exploration of the faraway red planet.

After a 3-billion-mile, nearly 10-year journey, on July 14, 2015, New Horizons became the first spacecraft to explore the dwarf planet Pluto's moon Charon.

QuickStart

Imagery appears throughout the ArcGIS platform. Here are two starting points.

▶ **High-resolution global basemaps**

Accessed billions of times monthly, the ArcGIS Imagery and Imagery with Labels basemaps are the most popular background maps that people use for their GIS projects. Imagery serves as a canvas to provide context and validation for your GIS data.

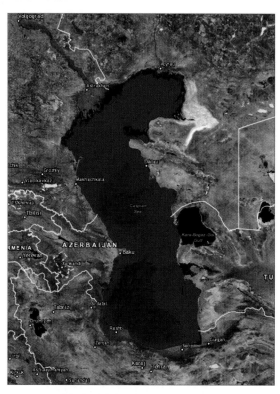

Imagery Basemaps in The Living Atlas

▶ **Landscape layers**

Landscape analysis underpins much of our land-use planning, how we manage natural resources and their relationship with the environment. The landscape layers in this group are configurable and provide access to hundreds of measures about aspects of the people, natural systems, and plants and animals that define the landscape of the United State and the rest of the world.

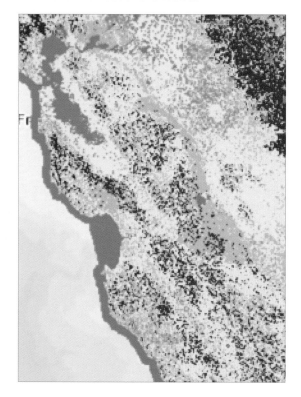

Landscape Layers in The Living Atlas

Learn ArcGIS lesson

Get Started with Imagery

In this lesson, you'll explore Landsat imagery and some of its uses with the Esri Landsat app. You'll first go to the Sundarbans mangrove forest in Bangladesh, where you'll see the forest in color infrared and track vegetation health and land cover. Then, you'll find water in the Taklamakan Desert and discover submerged islands in the Maldives. After using 40 years of archived Landsat imagery to track development of the Suez Canal over time, you'll be ready to explore the world on your own.

▸ Overview

Satellite imagery is an increasingly powerful tool for mapping and visualizing the world. No other method of imagery acquisition encompasses as much area in as little time. The longest-running satellite imagery program is Landsat, a joint initiative between two US government agencies. Its high-quality data appears in many wavelengths across the electromagnetic spectrum, emphasizing features otherwise invisible to the human eye and allowing a wide array of practical applications.

▸ Build skills in these areas:

- Navigating and exploring imagery
- Changing spectral bands to emphasize features
- Tracking changes over time
- Building your own band combination

▸ What you need:

- Estimated time: 15 to 30 minutes

Start Lesson

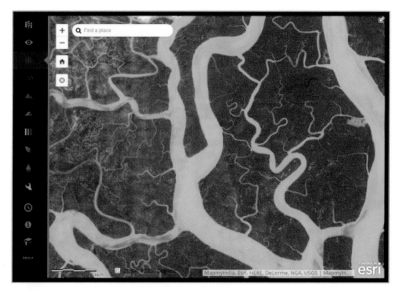

This app allows you to navigate the world with Landsat satellite imagery. Landsat takes images of the planet to reveal its secrets, from volcanic activity to urban sprawl. Landsat sees things in the electromagnetic spectrum, including what's invisible to the human eye. Different spectral bands yield insight about our precious and continually changing Earth.

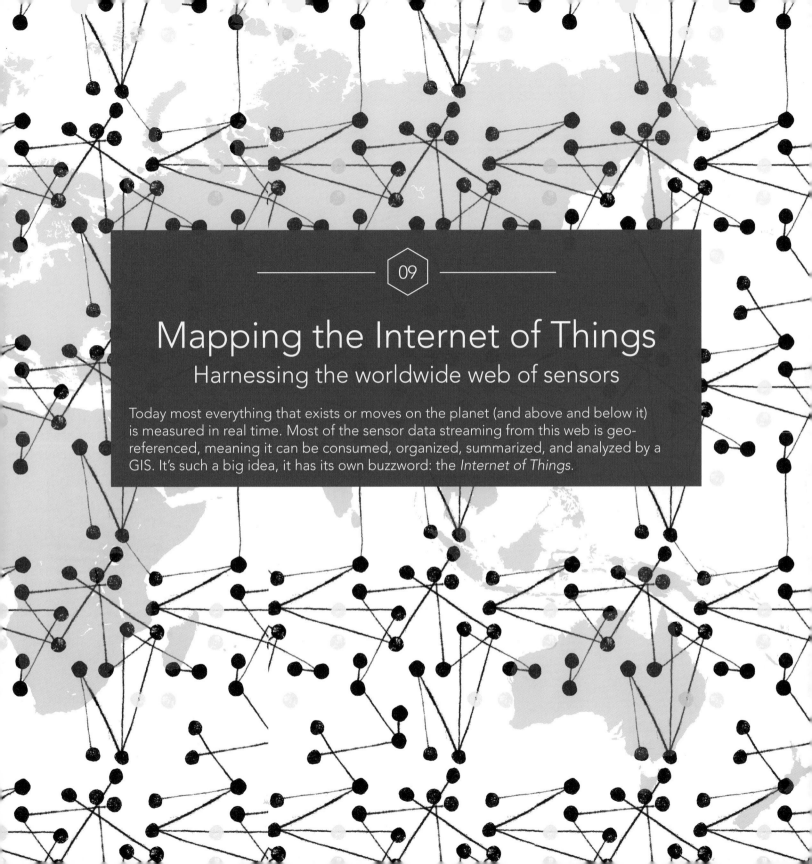

09

Mapping the Internet of Things
Harnessing the worldwide web of sensors

Today most everything that exists or moves on the planet (and above and below it) is measured in real time. Most of the sensor data streaming from this web is geo-referenced, meaning it can be consumed, organized, summarized, and analyzed by a GIS. It's such a big idea, it has its own buzzword: the *Internet of Things*.

The utility of real-time data in GIS

A vast amount of data is created every day from sensors and devices: GPS devices on vehicles, objects, and people; sensors monitoring the environment; live video feeds; speed sensors in roadways; social media feeds; and more, all connected through the Internet. What this Internet of Things means is that we have an emerging source of valuable data. It's called "real-time" data. Only recently has the technology emerged to enable this real-time data to be incorporated into GIS applications.

The real-time GIS capabilities of the ArcGIS platform have transformed how information is utilized during any given situation. Real-time dashboards fed by the IOT provide actionable views into the daily operations of organizations, empowering decision-makers and stakeholders with the latest information they need to drive current and future ideas and strategies. Dashboards answer questions such as: What's happening right now? Where is it happening? Who is affected? What assets are available? Where are my people?

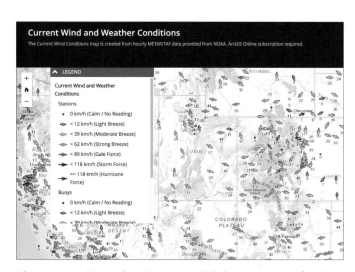

The National Weather Service publishes a series of real-time data feeds that can be readily consumed in ArcGIS and used to drive custom applications.

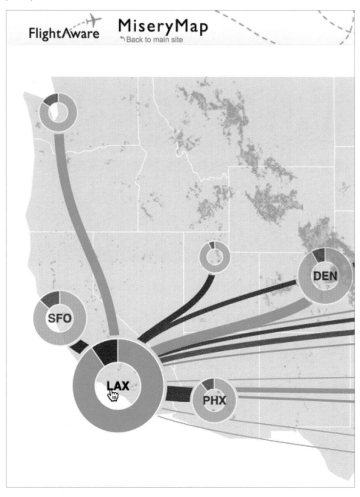

The FlightAware MiseryMap is a real-time visualization of the state of US flight delays and cancellations.

Some applications of real-time dashboards

- Local governments use real-time information to manage operations such as tracking and monitoring snowplows and trash trucks.

- Utilities monitor public services including water, wastewater, and electricity for consumers.

- Transportation departments track buses and trains and monitor traffic flows, road conditions, and incidents.

- Airport authorities and aviation agencies track and monitor air traffic worldwide.

- Oil and gas companies monitor equipment in the field, tanker cars, and field crews.

- Law enforcement agencies monitor crime as it happens, as well as incoming 911 calls.

- Companies use real-time social media feeds such as Twitter to gauge feedback and monitor social sentiment about particular issues.

- To issue early warnings and reports, federal agencies such as the Federal Emergency Management Agency (FEMA), US Geological Survey (USGS), National Oceanographic and Atmospheric Administration (NOAA), and Environmental Protection Agency (EPA) gather vast amounts of information about the environment. They monitor weather, air and water quality, floods, earthquakes, and wildfires.

- Individuals use elements of the IOT—smartphones, smartwatches, smart sensors, radio-frequency identifications (RFIDs), beacons, fitness bands, and so on—to capture and visualize information about every type of activity.

- Emergency management agencies monitor public safety during large events, such as marathons and the Olympics.

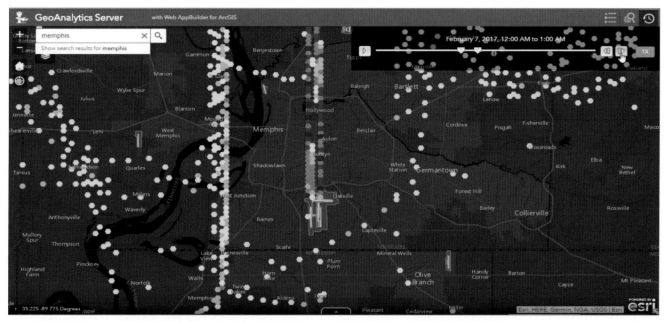

ArcGIS is used to monitor FedEx flight operations on a typical night at the Memphis International Airport.

How *real* is real time?

Real-time data is as current as the data source that is updating it, whether that data is being updated every second, minute, hour, or daily. What is real time to one organization might not be real time to another, depending on the type of scenario being monitored.

Real time is a concept that typically refers to the awareness of events at the same rate or at the same time as they unfold (without significant delay). It's often confused with frequency, or the intervals between events, which is essentially how often the event is updated. The update interval, or frequency, relates to the term "temporal resolution," which can vary from one application to another.

For example, most aircraft monitoring systems provide two updates every second, whereas it may take every hour to provide a weather update. For monitoring their networks, energy utilities use systems, also known as SCADA (Supervisory Control and Data Acquisition), that sample data about voltage, flow, pressure, and more from analog devices at very high frequencies (e.g., 50 hertz). This can result in high resource requirements for network bandwidth, system memory, and storage volume.

The data that fueled geographic applications in the past was created to represent the state of something at a specific point in time: data captured for what has happened, or what is happening, or what will happen. Although this GIS data is valuable for countless GIS applications and analyses, today the current snapshot of what is happening now falls out of sync very quickly with the real world, in many cases becoming outdated almost as soon as it is created.

What is real-time GIS?

Real-time GIS can be characterized as a continuous stream of events flowing from IoT sensors or data feeds. Each event represents the latest measured state, including position, temperature, concentration, pressure, voltage, water level, altitude, speed, distance, and directional information flowing from a sensor.

Maps provide the most basic frameworks for viewing, monitoring, and responding to real-time data feeds.

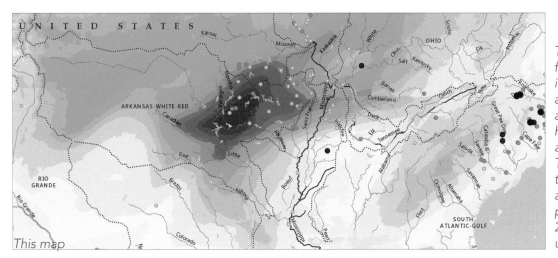

This national water map for the United States integrates weather feeds and storm warnings, along with real-time and historical stream gauges and weather forecasts. The map, updated several times daily, also includes a national water forecast predicting stream flows for 2.7 million river reaches for up to 10 days out.

Aspects of a real-time GIS

Acquire real-time data

A utility organization may want to visually represent the live status of its network with information that is captured by sensors in the field. Although the sensors on the network are not physically moving, their status and the information they send changes rapidly. Radio-frequency identification (RFID) is being used in a wide variety of environments to keep track of items of interest. Warehouses and logistics companies use RFID to track and monitor inventory levels. Hospitals use it to track equipment to make sure it has gone through proper cleansing procedures before being used.

A wide range of real-time data is accessible today. Connectors exist for many common devices and sensors enabling easy integration between the IoT and your GIS.

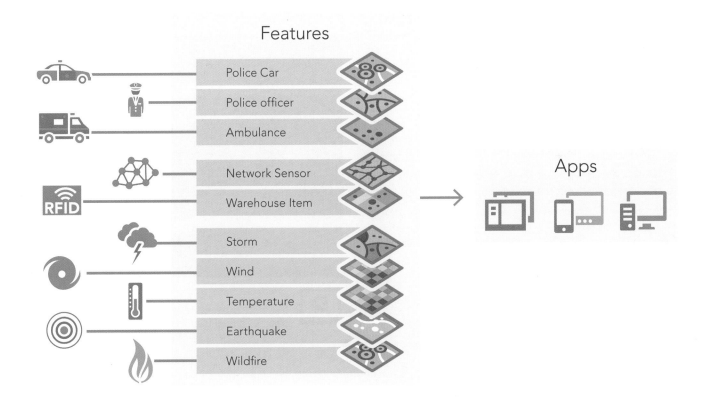

Thought leadership:
Suzanne Foss and Adam Mollenkopf

The Internet of Things is taking shape all around us. Large investments are being put into smart cities, autonomous vehicles, public safety services, utilities, and telecommunication infrastructure. The sensors that are being implemented are effectively digitizing our planet to a level we have never seen before. The sheer number of sensors, the abundant variety of sensor types available, and the frequency of updates that these sensors produce is a new opportunity for GIS communities to leverage the integration of Real-Time GIS and the IoT.

In many ways, the evolution of Real-Time GIS has been driven by IoT. Early systems focused primarily on automatic vehicle location and mobile asset monitoring. This has continued to develop and evolve as new types of sensors have become available and have dropped in cost.

Today, Real-Time GIS systems strongly complement IoT solutions by expanding capabilities to incorporate continuous space–time analysis. Autonomous vehicles are a good example in which vehicles report locations as well as observations about road conditions. These collective observations can be used in concert to analyze road conditions, and provide hazard alerts and alternative routing when required. The ability to combine information from multiple sensor types and locations is critical to managing complex operations.

This integration of different sensor networks is combined intelligently in a geospatial framework to optimize operation and is one of the biggest values of IoT. Previously disparate sets of information can be brought together, in real time, to see all facets of a problem and make smarter decisions, thus improving

efficiency, optimizing services, and reducing costs. And using a geospatial context is vital for this to be successful.

Geography is a natural integrator, and GIS systems play an important role in integrating the relationships between different sensor systems. The interaction between data streams and corresponding action are all essential in building smarter applications using real-time geoanalytics.

 Suzanne Foss and Adam Mollenkopf of Esri demonstrate advanced real-time geoanalytics at the 2016 Esri User Conference.

Components of a real-time dashboard

Real-time dashboards are created by adding "widgets" to an operation view. Operation views are easy to set up and configure. The map widget creates the primary map display and serves as the source of data for other widgets. You choose which data source or attribute value is displayed by the widget, specify the appearance settings, enter a description or explanatory text, and set any other properties required for the particular widget.

Widgets are used to represent your real-time data in a visual way. For example, a symbol could represent the location of a feature on a map; a text description could be displayed in a list; and a numerical value could be shown as a bar chart, gauge, or indicator.

Each operational view is updated with the latest data by setting a refresh interval on both the widget and each layer.

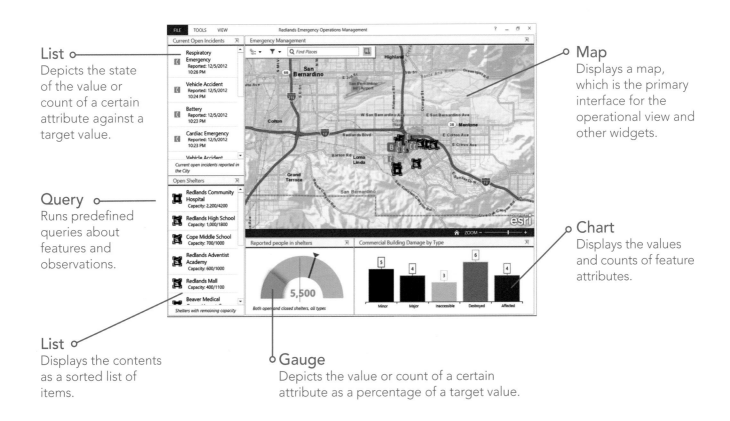

List
Depicts the state of the value or count of a certain attribute against a target value.

Query
Runs predefined queries about features and observations.

List
Displays the contents as a sorted list of items.

Map
Displays a map, which is the primary interface for the operational view and other widgets.

Chart
Displays the values and counts of feature attributes.

Gauge
Depicts the value or count of a certain attribute as a percentage of a target value.

Real-time GIS platform capabilities
Working with real-time data

ArcGIS® GeoEvent Server for ArcGIS® Enterprise is a GIS server extension that brings your real-time data to life, allowing you to connect to virtually any type of streaming data, process and analyze that data, and send updates and alerts when specified conditions occur, all in real time.

With GeoEvent Server, your everyday GIS applications become frontline decision applications, helping you respond faster with increased awareness whenever and wherever change occurs.

Connecting to feeds
GeoEvent Server is capable of receiving and interpreting real-time data from virtually any source. The system understands how the real-time data is being received as well as how the data is formatted. Input Connectors (shown here) allow you to acquire real-time data from a variety of sources.

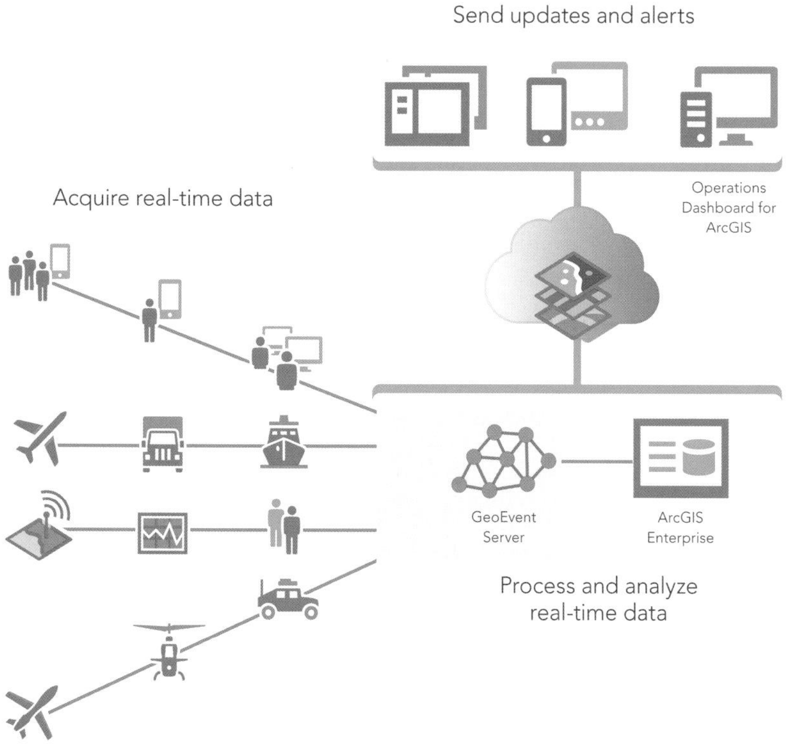

Sending updates and alerts
Output Connectors are responsible for preparing and sending the processed data to a consumer in an expected format. An Output Connector translates its events into a format capable of being sent over a particular communication channel.

Performing real-time analytics
GeoEvent services enable you to define the flow of the event data as well as to add any filtering and processing on the data as it flows to the Output Connector. Applying real-time analytics allows you to discover and focus on the most interesting and important events, locations, and thresholds for your operations.

Send updates and alerts

Acquire real-time data

Operations Dashboard for ArcGIS

GeoEvent Server

ArcGIS Enterprise

Process and analyze real-time data

Visualizing real-time data

With Operations Dashboard, you can create real-time dashboards that allow you to visualize and display key information about your operations. These operational views can be stored in ArcGIS and shared with individual members of your organization, with groups within your organization, and publicly with anyone using ArcGIS.

Real-time data storage

In many cases, data streamed into ArcGIS in real time will be captured in a geodatabase. To support historical archiving of events, a best practice is to use a historical or temporal feature class to store all the events received from the data. This allows the state of each object to be stored indefinitely, everything from the first event received until the present. As you can imagine, the size of this data can grow to be quite large, especially over an extended period of time. The growth rate of your data is largely dependent on the message size and the frequency of the incoming data. A best practice is to define and enforce a retention policy for how much history is actively maintained in the geodatabase.

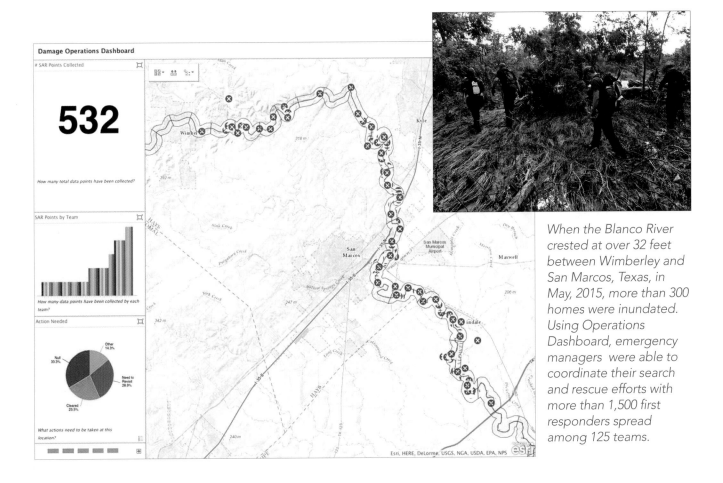

When the Blanco River crested at over 32 feet between Wimberley and San Marcos, Texas, in May, 2015, more than 300 homes were inundated. Using Operations Dashboard, emergency managers were able to coordinate their search and rescue efforts with more than 1,500 first responders spread among 125 teams.

Examples of real-time data sources

Real-time data takes on many different forms and has many different applications. Some of these examples link to live-feed maps and some to the item descriptions for the feeds themselves.

Active hurricanes

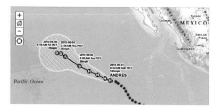

National Hurricane Center data describes the current and forecast paths of tropical activity.

Hourly wind conditions

The Current Wind Conditions layer is created from hourly data provided from NOAA.

USGS earthquakes

Minute-by-minute earthquake data for the last 90 days comes from the USGS and contributing networks.

LA Metro bus locations

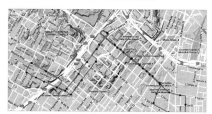

LA Metro's Realtime API gives access to the positions of Metro vehicles on their routes in real time.

Stream gauges

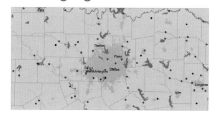

These stream gauge feeds allow users to map current water levels to monitor flood and drought risk.

World traffic

Updated every five minutes, this dynamic map service monitors traffic speeds and incidents.

Twitter feeds

ArcGIS provides a sample for displaying geolocated tweets live on a web map.

Instagram feeds

ArcGIS provides a sample for displaying geolocated Instagram posts live on a web map.

Severe weather

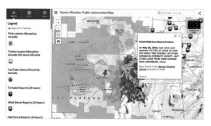

This map features live-feed layers for severe weather across the United States and Canada.

Case study: Real-time flood warning
North Carolina Floodplain Mapping Program

North Carolina established the North Carolina Floodplain Mapping Program (NCFMP) to better identify, communicate, and manage risks from flood hazards within the state in response to the devastating flooding caused by Hurricane Floyd in 1999. This led to the establishment of the Flood Inundation Mapping and Alert Network (FIMAN) to provide real-time flood information throughout the state.

In the first week of October 2015, the system was put to the test through the combination of Hurricane Joaquin passing to the east and a stalled low-pressure system that produced historic rainfall totals and subsequent flooding in portions of the Carolinas. The storm, which had rainfall totals ranging from three inches to more than 20 inches over a three-day period, resulted in more than 20 fatalities and damages estimated in the billions of dollars.

Although North Carolina was spared the extreme rainfall experienced in South Carolina, the storm still resulted in significant flooding along the coast and eastern counties. FIMAN was used by the State Emergency Operations Center throughout the storm to monitor

Flooding in Edgecomb County, North Carolina.

flooding conditions, assess potential impacts of flooding on the basis of weather forecasts, and target the deployment of emergency response personnel and resources. FIMAN served as an invaluable tool in communicating risk to public officials and the public.

Visualization of impacted buildings mapped in near real time with the actual flooding.

QuickStart

Get your real-time dashboard up and running

Operations Dashboard for ArcGIS is a Windows app that you can download and run locally or a web-based version that runs in a browser. It's where you design your operational views.

1. Download and install Operations Dashboard for ArcGIS
2. Documentation is available online.
3. An ArcGIS organizational account is required.

Tips for real-time dashboards

There are many principles to consider when configuring real-time dashboards:

- Design it for a specific purpose or scenario.
- Keep it easy to understand and intuitive, so no one needs to ask for explanations.
- Make the layout simple so that it focuses attention on the most important information.
- Present the information in a prioritized way that assists in making timely decisions.
- Render it flexible enough to be delved into for more detail when necessary.
- Make sure it provides timely updates and synchronizes all the widgets in real time.

Multidisplay versus single-display dashboards

Operations Dashboard for ArcGIS provides two kinds of operations views:

- Multidisplay operations views are useful in environments with multiple monitors (as in a desktop setting). They are especially useful when you have a centralized operations center where staff are collectively viewing multiple monitors displaying continually updated maps, charts, and video feeds.
- Single-display operations views are designed for individuals to use on mobile phones, tablets, and web browsers.

GeoEvent Server

This extends ArcGIS Enterprise and provides capabilities for consuming real-time data feeds from a variety of sources, continuously processes and analyzes that data in real time, and updates and alerts your stakeholders when specified conditions occur.

To learn more about GeoEvent Server, access the documentation, sample connectors, and videos at links.esri.com/geoevent.

Learn ArcGIS lesson

Create a Real-Time Layer

Overview

You work for the City of Sandy Public Works Department and would like to track progress of snow removal operations. The progress will be rendered on the map to show the status of each street in the city. The map will be used by both city personnel and citizens so everyone can see in real time whether the street of interest is passable or not.

In this series of lessons, you'll track the location of the Sandy Public Works snowplow fleet and the status of the main arterial of the Sandy street network. First, you will learn how to get the real-time locations of the Public Works vehicles into the ArcGIS platform, and you'll be aware that vehicles may or may not be snowplows. You'll author the logic to process the vehicle data instantly at the time of receiving such data. The location of snowplow vehicles will be used to update the status of the street. In this case, assume that if a snowplow is moving on a street, it is working on removing snow to clear the street.

▸ Build skills in these areas:
- Configuring GeoEvent connectors
- Designing and authoring real-time processing logic
- Filtering for snowplow vehicles
- Symbolizing the current location and vehicle tracks of each vehicle

What you need:

- Publisher or administrator role in an ArcGIS organization
- Access to GeoEvent Server
- Estimated time: one to two hours

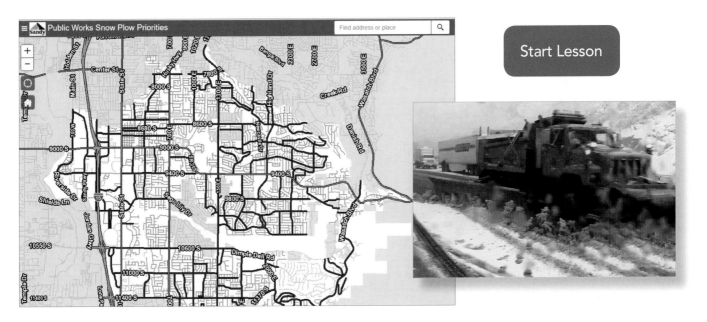

GIS Is about Community
Web GIS is the GIS of the world

Your own GIS is simply your view into the larger system. It's a two-way street. You consume information that you need from others, and in turn, you and others feed your information back into the larger ecosystem.

GIS work is a valued profession
Community is vital in GIS

It is important to recognize the phenomenal growth of GIS in people's lives and how its effect extends beyond its economic and fiscal impacts. You are—and should be—an active participant in a truly amazing field. Every day, millions of people are using GIS in government, industry, and academia. Even smaller organizations are hiring dedicated GIS professionals to improve the quality and accuracy of work being accomplished, and the benefits of doing so are immeasurable. GIS helps people make better decisions, reduce costs, work more efficiently, communicate better, and gain key insights.

Globally, GIS and the related geospatial economy is valued at more than $250 billion per year. The geospatial segment is one of the fastest growing

Gathering of the faithful at an Esri user conference. The UC has been an annual event since 1981.

in the tech field overall. And that's saying something, because everyone knows how fast tech is growing. This segment is considered by the US Department of Labor to be one of the three technology areas that will create the greatest number of new jobs over the coming decade. It's growing at 35 percent overall, and some of the sectors such as business GIS are growing at a 100 percent clip.

This worldwide community is busy every day implementing GIS: It is growing its expertise and extending its reach throughout organizations and communities all over the planet. The work these people do is impossible to pigeonhole because it is so broad, yet much of it tends to focus its attention on critical resource issues, environmental management and mitigation, climate change, key urban initiatives, and other daunting problems. The most committed users of GIS tend to be passionate and interested in the world, and dedicated to making a difference. To them, and maybe to you, it's important to feel like the work you do means something.

GIS is also face-to-face

For a lot of the reasons described in this book, GIS is a collegial profession with a strong networking aspect. Organizations including the Urban and Regional Information Systems Association (URISA), American Association of Geographers (AAG), and others have long held well-attended conferences. From the beginning, Esri has encouraged face-to-face networking of its user community through regional user groups, industry user groups, its Developer Summit, and its annual Esri User Conference. With more than 17,000 in attendance, this event is the largest GIS gathering annually. To sit in the audience during the plenary session is to feel part of something truly larger than yourself.

GIS is collaborative
Geography is key to integrating work across communities

Modern GIS is about participation, sharing, and collaboration. As a Web GIS user, you require helpful, ready-to-use information that can be put to work quickly and easily. The GIS user community fulfills that need—that's the big idea in this chapter. GIS was actually about open data long before the term gained fashion because the people who were doing it were always looking for ways to deepen and broaden their own GIS data holdings. No one agency, team, or individual user could possibly hope to compile all the themes and geographic extents of data required, so people networked and shared to get what they needed.

Since the early days in GIS, people realized that to be successful, they would need data from other sources beyond their immediate workgroups. People quickly recognized the need for data sharing. Open GIS and data sharing gained traction quite rapidly across the GIS community, and continues to be a critical aspect of GIS implementation today. With cloud computing and the mobile/app revolution, the GIS community is expanding to include almost everyone on the planet. The data in every GIS is being brought together virtually to create a comprehensive GIS of the world, and nearly everyone can take GIS with them everywhere they go on their tablets and smartphones. Geography and maps enable all kinds of conversations and working relationships both inside and outside your organization.

GIS is for organizations—First and foremost, your GIS can be used by people throughout your organization. In Web GIS, maps are purpose-driven, and their intended audience may include executives, managers, decision-makers, operations staff, field crews, and constituents. ArcGIS Online enables you to extend your reach to these users.

Clint Brown: GIS is a social activity

GIS is for communities—GIS users collaborate across communities. These communities may be based on relationships fostered by living in the same geography (a city, region, state, or country) or by working in the same industry or subject matter (conservation, utilities, government, land management, agriculture, epidemiology, business, and more). In these communities, users share critical data layers as well as map designs, best practices, and GIS methods.

GIS is for public engagement—People everywhere are starting to engage with GIS. They have been using maps as consumers, and now they are interested in applying them at work and in their community relationships. Often this involves communicating with the public by telling stories using maps. More and more, members of the public are providing input and collecting their own data for GIS organizations and the public good. This sharing of data makes for better civic engagement at multiple levels.

ArcGIS is for organizations

GIS has a critical role to play in your organization. ArcGIS is a platform that enables you to create, organize, and share geographic information in the form of maps and apps with workers throughout your organization. These run virtually anywhere—on your local network or hosted in the ArcGIS Online cloud. The maps and apps that you share are accessible from desktops, web browsers, smartphones, and tablets.

The role of the GIS department

Professional GIS provides the foundation for GIS use across your organization. It all starts with the work you do on your professional GIS desktops. You compile and manage geographic data, work with advanced maps, perform spatial analysis, and conduct GIS projects. Your resulting GIS content can be put to use by others in countless ways. Your work is shared as online maps and apps that bring GIS to life for users within your organization and beyond.

Portals enable collaboration across your organization

A key component in your organizational GIS is your information catalog or portal. This catalog contains all the items (maps, layers, analytical models, apps) that are created, used, and ultimately shared by your group's users.

Every item is referenced in your organization's information catalog—your portal. Each item contains a description (often referred to as metadata), and any item can be shared with selected users within as well as outside your organization.

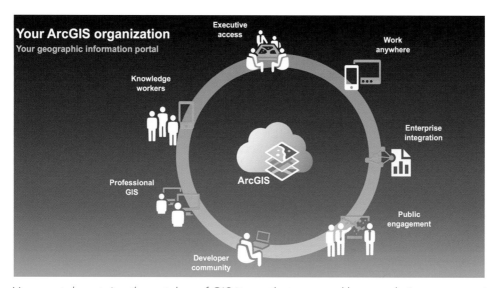

Your portal contains the catalog of GIS items that are used by people in your organization. These items include maps, scenes, layers, analytical models, and apps.

ArcGIS provides intelligent online content management that enables you to create and share useful maps and apps with your users. You can engage ArcGIS to organize and distribute your geographic information and tools. With a portal, certain users will have access to apps that support their specific work tasks, such as apps configured for collecting data in the field. Some maps will be shared with the entire organization, such as the basemaps that provide the foundation for all the work performed across your organization. Some users will create their own maps by mashing up their data layers with those of others. And some items, such as story maps about your organization's work, might be shared with everyone including the public.

Access to GIS content across your organization

1. Start with your GIS content and your organization's GIS content.
2. Combine it with community information layers shared by users with whom you collaborate, along with layers shared by the broader GIS community.
3. Create maps and analytical tools for your users and constituents, and share these online.
4. Share your maps and geographic information layers with others throughout your organization and, optionally, beyond.

GIS roles

GIS is about the people in your organization and the purpose-driven maps and apps they apply to do their work. Every user is given an ArcGIS account (i.e., a login) and assigned a role in using ArcGIS. For example:

- **Administrators** manage the system and enable new users to participate by granting privileges for their roles in your organization's GIS. There are only a handful of administrators (one or two) in each organization.
- **Publishers** create maps and apps that can then be shared with users throughout your organization and the public. Publishers also help organize content by creating and managing logical collections or groups. Users throughout your organization find their maps and apps in these logical groups.
- **Users** create and use maps and apps, and then share them with others—inside and outside your organization.
- **Viewers** use maps and apps and perform basic operations such as searching and geocoding.

Building smarter communities
An initiative-driven approach to community engagement

Communities across the globe today are facing many specific issues, and ArcGIS is a well-established and reliable tool used to help address these issues.

The ArcGIS Hub offers a new approach to tackling these problems that brings together executives and staff, nonprofit organizations, and citizens in a common framework. An initiative is defined—such as reducing crime—and the ArcGIS Hub steps you through a process of finding foundational data, deploying apps, engaging the community, collecting more data, performing analysis, and measuring actions and results. The ArcGIS Hub makes it easy to set up and run an initiative, and multiple initiatives can be active in your community at any given time. Three example initiatives are highlighted here: reducing vector-borne disease, reducing homelessness, and reducing opioid addiction.

Reduce Vector-borne Disease
Smarter communities don't just react to pest emergencies; they work hard to prevent them. An ArcGIS-based, initiative-driven approach provides your community with the intelligence you need to understand and mitigate vectors in your community. Field crews and decision-makers are empowered to improve prevention, mitigation, response time, and public engagement.

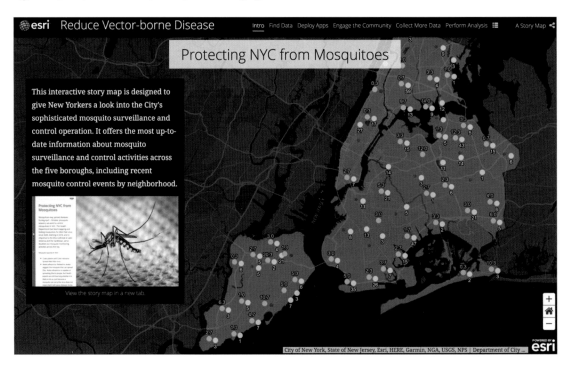

Reduce Homelessness

As homelessness increases because of economic and social factors, communities are working hard to eliminate the problem. Empowered with GIS, government staff, nonprofit workers, and volunteers scour the streets to survey the people living without shelter. This information can then be aggregated to identify where homeless individuals and encampments are located. Once completed, spatial analysis tools can be employed to help target and deliver services to impacted populations.

Reduce Opioid Addiction

Communities are being pulled apart by opioid abuse and overdoses, and governments and citizens need to respond quickly as the problem continues to grow. The ability to see through the emotion and make data-driven decisions is critical. An initiative-driven approach allows you to get a visual on the areas that need help the most, engage with your organization's other departments in new ways, and work with the public to drive awareness and help eliminate the abuse.

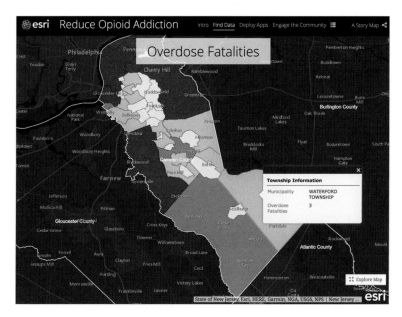

Other initiatives

Countless other initiatives can be implemented, including Vision Zero (eliminating pedestrian and cyclist deaths), improving walkability, increasing affordable housing, reducing crime, increasing public trust and police transparency, reducing traffic, improving disaster preparation, ensuring accessibility for the disabled, protecting critical infrastructure, creating safe routes to school, and many more. Some of these are included "out-of-the-box" with the ArcGIS Hub, while others can easily be custom built to meet your specific needs. With the ArcGIS Hub, the number and types of initiatives you tackle is limited only by the needs and imagination of your community.

Geodesign

Using social engagement in community planning

"Everyone designs who devises courses of action aimed at changing existing situations into preferred ones."

—Herbert Simon,
political scientist (1916–2001)

Geodesign provides a planning methodology and approach to project design and decision-making, and it is best practiced by a community of collaborators. A technical design approach is also involved. Once objectives for a project are articulated, professionals survey and characterize a landscape. They identify its special resources and the opportunities to support a project as well as the constraints that limit what might be possible or practical. GIS is often used in this phase to perform suitability and capability analysis. These results are used to generate the landscape of opportunities and constraints. Subsequently, design alternatives are sketched onto the landscape, and further GIS analysis is used to evaluate, compare, and analyze the various design alternatives.

The practice of geodesign requires collaboration among project participants. The most important aspect is the feedback and ideas that are generated by the participants—including local citizens and stakeholders who may be affected. Most geodesign activities are about this kind of community engagement and consideration. GIS provides a useful tool for others to participate in the evaluation by providing the ability to consider the issues of other stakeholders.

Many problems in the world are not well defined, not easily analyzed, and not easily solved. What we do know is that the issues are important and require thoughtful consideration. They are beyond the scope and knowledge of any one person, discipline, or method. People must begin to understand the complexities, and then figure out ways to collaborate. Collaboration is a common thread, and social benefits are the central objective.

Geodesign, as an idea, has the potential to enable more effective collaboration between the geographically oriented sciences and the multiple design professions. It is clear that for serious societal and environmental issues, designing for change cannot be a solitary activity. Inevitably, it is a social endeavor.

—Adapted from *A Framework for Geodesign: Changing Geography by Design* by Carl Steinitz

 In this TED Talk, Esri president Jack Dangermond speaks about geodesign, a concept that enables architects, urban planners, and others to harness the power of GIS to design with nature, geography, and community in mind.

Thought leader: Lauren Bennett
Spatial analysis is changing everything

Across almost every industry and discipline, we are seeing an unprecedented focus on analytics. Never have I been so excited to see the wide range of ways that organizations are using spatial analysis. From crime analysis to disease surveillance to retail and public policy, it seems that the world is awakening to the power of thinking spatially. As data science becomes pervasive, the science of where is taking on a central role in the way that organizations think about their data and make informed decisions.

One area that I'm particularly passionate about is spatiotemporal analysis, where the world's data is finally reaching its full potential as it is visualized, analyzed, and transformed into powerful insight by taking advantage of both the spatial and time characteristics intrinsic in most data.

We're seeing large commercial organizations and small environmental firms analyze everything from customer sales to deforestation, and they are revealing information that has been hidden in their data, in many cases, for decades – enabling their organizations to understand trends and anticipate the future.

And probably the most exciting thing is that spatial analysis isn't just being used by experienced GIS professionals (though certainly this community is at the forefront!). Spatial reasoning is finding its way into the hands of all sorts of analysts and knowledge workers who are posing new questions and getting excited about this new way of thinking about their data. The Science of Where is everywhere, and it may just change everything.

Lauren Bennett is a spatial statistics specialist at Esri where she helps develop cutting-edge GIS software and methods for advanced geospatial analysis.

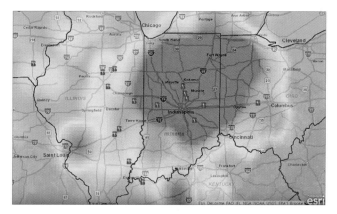

Precision Agriculture with GIS imagery by Beck's Hybrids

Social GIS and citizen science

Always looking for compelling and useful ways to engage with their local communities, many organizations are discovering that geography and maps are the perfect way to facilitate civic engagement and citizen science.

Tree survey

This citizen science demonstration project shows how volunteers can use a storytelling app to contribute to a citizen science project to identify the locations and conditions of trees in their community.

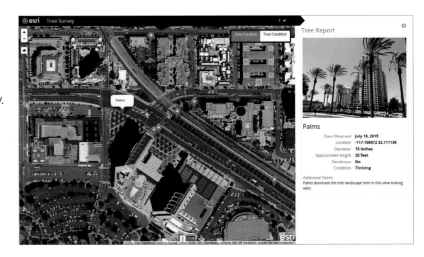

Crowdsource apps

The Crowdsource Reporter app allows citizens to submit problems or observations within their community.

The Crowdsource Manager app allows users within an organization to review problems or observations submitted through the Reporter application.

The Crowdsource Polling app lets constituents submit comments or feedback on existing plans and proposals.

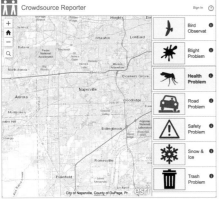
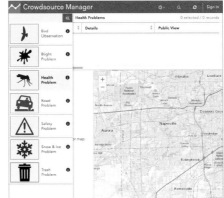
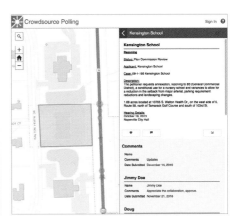

GIS is for public engagement

Citizens and constituents everywhere are beginning to embrace GIS and engage with local organizations. Here are a few examples of community engagement.

Cumberland Colours

This crowdsource story map created by the municipality of Cumberland in Nova Scotia, Canada, encourages public engagement through the sharing of favorite spots that show off fall colors across Cumberland County.

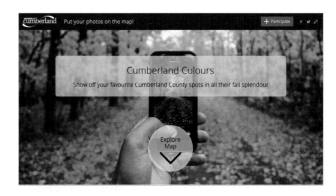

1Frame4Nature

We are all connected to nature, and through it, to each other. The International League of Conservation Photographers, a US-based, non-profit organization whose mission is to further environmental and cultural conservation through ethical photography, created a crowdsource story map through which photographers worldwide can share their pictures and stories of their personal connection to nature.

Downtown Eugene Is Happening

This crowdsource story map created by the City of Eugene, Oregon, lets visitors and residents explore the map to see great places downtown as well as submit something awesome they saw themselves.

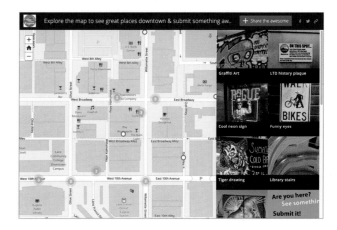

QuickStart

Participate! Share your maps, apps, and data with others

▸ **Use ArcGIS Open Data to share datasets.**
You can share your open data for use by everyone, without restrictions in minutes with the ArcGIS Open Data application. Organizations worldwide are opening up and sharing selected datasets with the public, enabling others to leverage their deep investments in critical information.

▸ **Create your own story maps for "the crowd"**
Publish a crowdsourced story map in which many people can participate and contribute. Use these to encourage community participation. Engage your community and encourage participation using story maps like this one on *Why Science Matters*.

▸ **Contribute your information to the Living Atlas of the World and to Community Maps.**
It's no secret: we all need information from each other's GIS. Our GIS work becomes stronger through this sharing. One of the most effective ways to share is by contributing your content to the Living Atlas and to the Community Maps for ArcGIS. Thousands of contributors have shared their best maps and GIS data with the world through the Living Atlas and amazing contributions to the Community Maps database.

▸ **Share GIS in your community and schools**
GIS users everywhere are engaging with people in their local communities through various types of community engagements. Many of these apply GIS and encourage its use and adoption. It's easy to contribute here.

▸ **Share this book.**
The "digital twin" of this book exists online as a free digital publication at www.thearcgisbook.com. This website contains literally hundreds of ready-to-use, live examples of ArcGIS in action, and makes it easy to get anyone from 8 to 80 up and running with ArcGIS and eager to go further. This is also a great way to share the idea of GIS with your friends, family, and co-workers.

Instructor's Guide for this book
Two world-renowned GIS educators in K12, Kathryn Keranen and Lyn Malone, have written a companion guide for use of this book in classrooms—from K12 through college that provide amazing resources and guidance for going further with GIS.

Participate in GIS Day.
Every November, there is a global event called GIS Day where GIS groups worldwide open their offices and classrooms for GIS Day. GIS Day provides an international forum for users of geographic information systems (GIS) technology to demonstrate real-world applications that are making a difference in our society.

Learn ArcGIS lesson

Set Up an ArcGIS Organization

You're an instructor at Laurel Junction, a community college in central Pennsylvania. The Geography Department is considering using ArcGIS Online to help teach students how to analyze data with maps. As a member of the department, you've been tasked with setting up a trial ArcGIS organization so you and other instructors can evaluate whether it would be a good resource for your courses.

▶ Overview

A colleague familiar with administering ArcGIS Online advised you on the initial steps to set up a trial organization. First, you'll activate the trial and complete some basic configuration tasks. You want an appealing site, so you'll add a custom banner and feature some apps and maps on the home page. You'll review some calculations to understand how credits are charged for the tools and storage that your department will use. You'll also create accounts for four instructors who will help you test. Finally, you'll learn where to download ArcGIS Pro, ArcGIS® Maps for Office®, and other apps, and assign licenses to members. Once your organization is ready for use and you are familiar with basic administration tasks, you can continue on your own with a more thorough configuration of the site.

Build skills in these areas:

- Administering an ArcGIS organization
- Designing the home page
- Sharing content and creating groups
- Creating custom roles and adding members
- Calculating credits for analysis and storage
- Managing licenses for ArcGIS apps

▶ What you need:

- Administrator role in an ArcGIS organization
- Estimated time: 15 to 30 minutes

Start Lesson

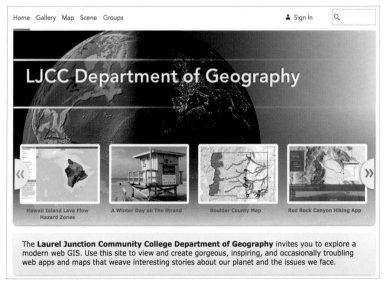

Home   Gallery   Map   Scene   Groups   ▲ Sign In   🔍

LJCC Department of Geography

Hawaii Island Lava Flow Hazard Zones   A Winter Day on The Strand   Boulder County Map   Red Rock Canyon Hiking App

The **Laurel Junction Community College Department of Geography** invites you to explore a modern web GIS. Use this site to view and create gorgeous, inspiring, and occasionally troubling web apps and maps that weave interesting stories about our planet and the issues we face.

Join a thriving community

More than 350,000 organizations—and over 3 million people—use Esri software to create the maps that run the world. When you start using Esri software, you join a thriving community of people and organizations who are excited to share the innovative and important work they are doing.

GeoNet

GeoNet is where the Esri community—customers, partners, Esri staff, and others in the GIS and geospatial professional community—connect, collaborate, and share experiences. GeoNet is the central online destination where the Esri community gathers to exchange ideas, solve problems, accelerate success, and build relationships to create a better world through the use of geographic information technology. Visit geonet.esri.com.

Esri partners

More than 2,000 value-added resellers, developers, consultants, data providers, and instructors partner with Esri to deliver a variety of value-added services and solutions to the GIS user community worldwide. For a complete list and description of partners' offerings, visit www.esri.com/partners.

ArcGIS Marketplace

ArcGIS Marketplace is your destination to search, discover, and get apps and content provided by Esri partners, distributors, and Esri. Apps listed in the marketplace are built to leverage and enhance what your organization can do with GIS. The marketplace includes both paid and free apps, and many include the ability to access free trials. Visit marketplace.arcgis.com.

Conferences

Esri hosts dozens of user conferences around the world each year. The largest of these, the Esri User Conference, is a gathering of more than 16,000 GIS professionals held each summer in San Diego, California. User conferences offer many hours of hands-on training, moderated sessions, technical workshops and demonstrations, user presentations, inspirational speakers, and much more. Visit esri.com/events.

Offices worldwide

Esri has more than 80 distributors around the world and 10 regional offices in the United States. Visit this interactive map and click your location to find someone who can help you.

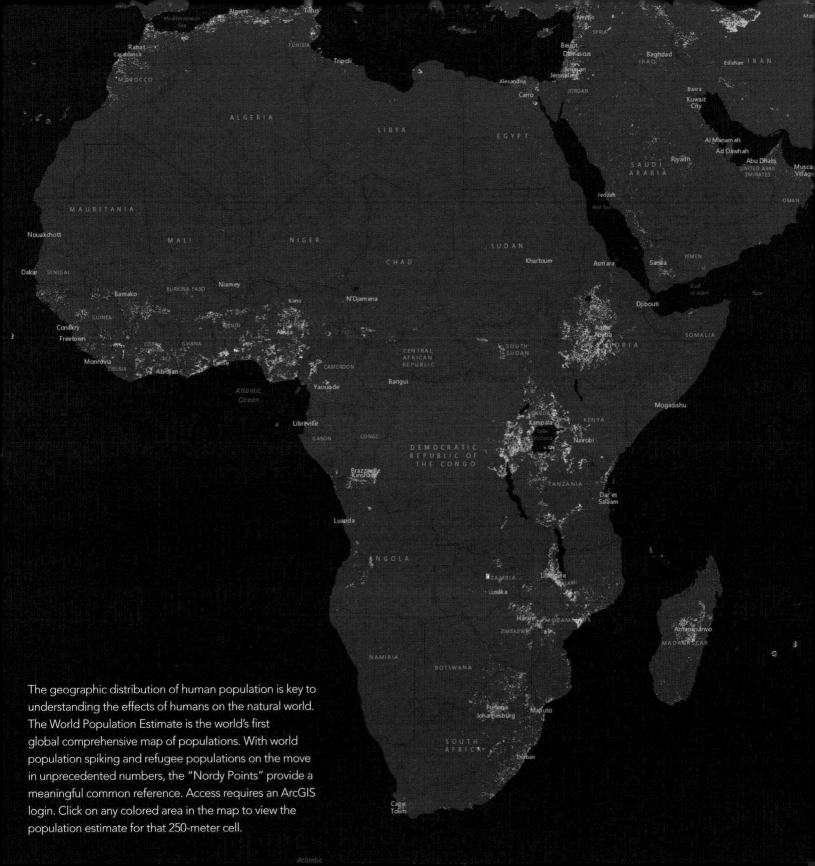

The geographic distribution of human population is key to
understanding the effects of humans on the natural world.
The World Population Estimate is the world's first
global comprehensive map of populations. With world
population spiking and refugee populations on the move
in unprecedented numbers, the "Nordy Points" provide a
meaningful common reference. Access requires an ArcGIS
login. Click on any colored area in the map to view the
population estimate for that 250-meter cell.

Learn ArcGIS gallery

Learn ArcGIS is a gallery of real-world problem-solving lessons.

Get Started with ArcGIS Maps for Office
Use spreadsheet data to find potential customers near a new salon.

Research Market Potential Using Esri Business Analyst Web App
Determine whether downtown Des Moines is ready for an uptown movie house.

Analyze Crime Using Statistics and the R-ArcGIS Bridge
Perform statistical analysis of San Francisco crime using the R ArcGIS bridge.

Get Started with Survey123 for ArcGIS
Survey safety prevention measures and create an inventory of emergency assets.

Find Areas at Risk of Flooding in a Cloudburst
Use ModelBuilder to analyze drainage

Plan Routes for Food Inspectors
Help four health officials inspect 36 restaurants in San Diego County.

Get Started with ArcMap
Map the impact of roads on deforestation in the Amazon rainforest.

Monitor Real-Time Emergencies
Keep track of fast-changing situations with Operations Dashboard for ArcGIS.

Tell the Story of Irish Public History
Gain historical insight by mapping fatalities from Ireland's 1916 Easter Rising.

Set Up an ArcGIS Organization
Configure the site for a new ArcGIS organization.

Get Started with the Scene Viewer
Create a 3D scene showing Florida's beaches and inlets.

Map Voter Data to Plan Your Campaign
Identify political advantages to help your candidate win the election.

Oso Mudslide - Before and After
Show disaster imagery by creating an app with Web AppBuilder.

Manage a Mobile Workforce
Collect fire hydrant inspection results from the field using Collector for ArcGIS.

Evaluate Locations for Mixed-Use Development
Find the best areas for a new mixed-use housing development.

Hiking Red Rock Canyon
Educate hikers about trail difficulty with landscape layers from the Living Atlas of the World.

Fly Through South America in a 3D Animation
Animate a 3D tour of a famous geographer's epic journey.

Get Started with Tapestry
Use ZIP Codes to learn how Americans think, feel, and live.

Calculate Impervious Surfaces from Spectral Imagery
Classify land use types in an image to find impervious surfaces.

Track Crime Patterns to Aid Law Enforcement
Help the Lincoln Police Department allocate resources to combat crime.

I Can See for Miles and Miles
Identify areas from which turbines on a proposed wind farm would be visible.

Streamline Deliveries with Drive-Time Analysis
Create delivery zones so Wok & Roll stays in business.

No Dumping - Drains to Ocean
Learn about finding upstream watersheds and downstream flow paths from point locations.

Get Started with ArcGIS Pro
Create 2D and 3D maps to analyze flooding in Venice, Italy.

Get Started with Drone2Map for ArcGIS
Transform drone imagery into 3D GIS data.

Get Started with ArcGIS Earth
Navigate a 3D world, add data from online, and share your results.

Survey Customers to Gain Marketing Insight
Build a customer survey for a technology store franchise.

Extract Roof Forms for Municipal Development
Create realistic 3D roof forms from lidar data.

Homeless in the Badlands
Examine North Dakota's homeless problem by mapping federal data.

From London to Tokyo
Use the Urban Observatory to compare cities and ask spatial questions.

Mapping the Public Garden
Build a problem-alert web app for a community garden.

A Place to Play
Find sites for a new park near the Los Angeles River.

Depict Land-Use Change with Time-Enabled Apps
Use historical imagery and time animation to show land-use change in Thailand.

Fight Child Poverty with Demographic Analysis
Locate children in poverty using demographic analysis, smart mapping, and a web app.

Download Imagery from an Online Database
Search Landsat databases for multispectral imagery of Singapore.

Actionable Intelligence
Identify the information needed to stop insurgent missile attacks on your base.

Identify Landslide Risk Areas in Colorado
Analyze soil maps to predict future mud flows in rain-soaked Colorado.

Get Started with ArcGIS Online
Learn the basics of making maps online.

The Power of Maps
Try some of the many ways people use ArcGIS maps.

Make a GeoPortfolio
Create a head-turning GeoPortfolio of your work.

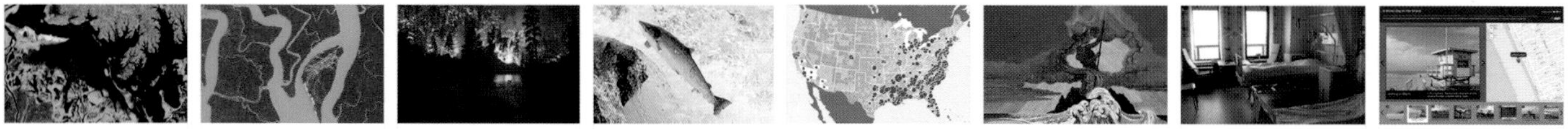

Classify Land Cover to Measure Shrinking Lakes
Compare imagery to calculate area change in Lake Poyang, China.

Get Started with Imagery
Explore 40 years of Landsat imagery from around the world.

Assess Burn Scars with Satellite Imagery
Calculate a burn index using imagery bands to measure fires in Montana.

Connect Streams for Salmon Migration
Propose a location for fishway construction and quantify the accessible habitat.

Bridging the Breast Cancer Divide
Learn about the disparity in breast cancer mortality rates.

Analyze Volcano Shelter Access in Hawaii
Analyze emergency shelter access on the island of Hawaii.

Where Does Healthcare Cost the Most?
Find hot spots in the cost of United States medical care.

Get Started with Story Maps
Create a photo map tour with a smartphone and Flickr account.

Further reading, books from Esri Press

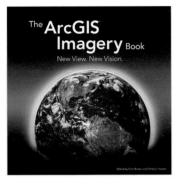

The ArcGIS Imagery Book
By Clint Brown and Christian Harder
ISBN: 9781589484627

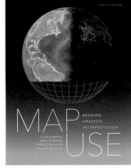

Map Use, 8th edition by Kimerling,
Buckley, Muehrcke, and Muehrcke
ISBN: 9781589484429

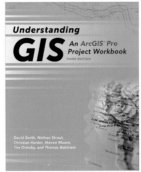

Understanding GIS, 3rd edition
By Smith, Strout, Harder, Moore,
Ormsby, and Balstrøm.
ISBN: 9781589484832

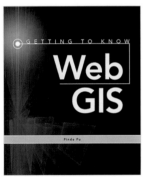

Getting to Know Web GIS,
2nd edition By Pinde Fu
ISBN: 9781589484634

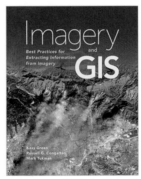

Imagery and GIS by Kass Green,
Russell G. Congalton, and Mark
Tukman
ISBN: 9781589484542

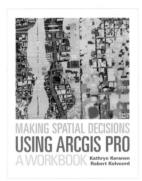

Making Spatial Decisions Using
ArcGIS Pro By Kathryn Keranen and
Robert Kolvoord
ISBN: 9781589484849

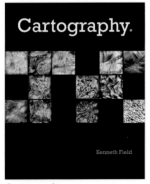

Cartography.
By Kenneth Field
ISBN: 9781589484399

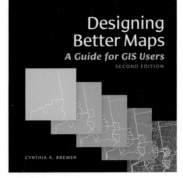

Designing Better Maps, 2nd edition
By Cynthia A. Brewer
ISBN: 9781589484405

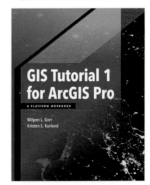

GIS Tutorial 1 for ArcGIS Pro
By Wilpen L. Gorr and
Kristen S. Kurland
ISBN: 9781589484665

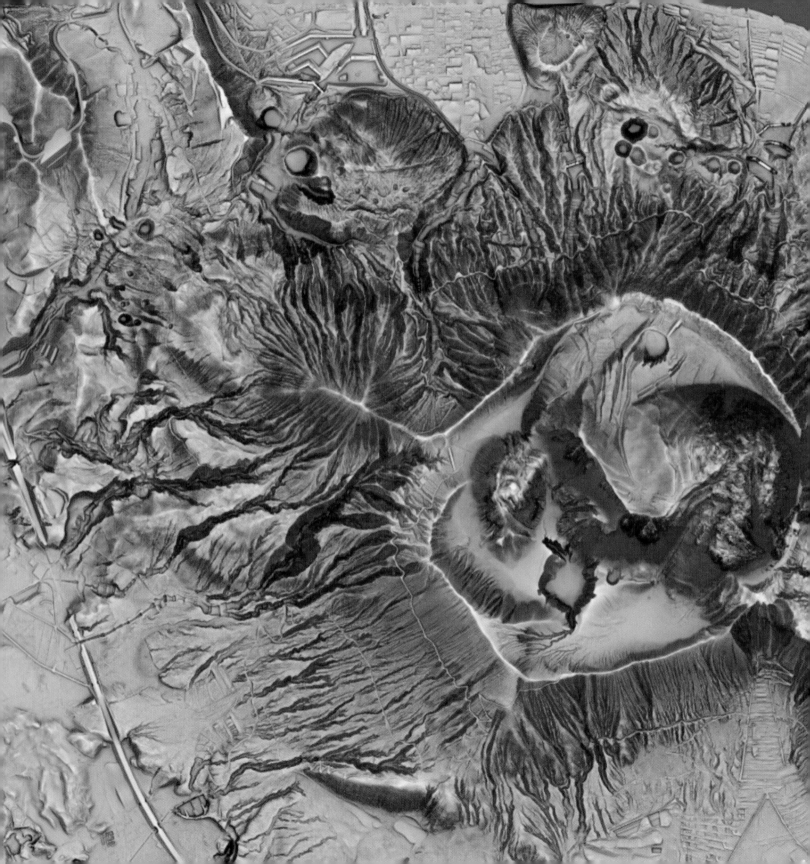

Contributors and acknowledgments

Contributors

Volume Editors: Christian Harder and Clint Brown
Chapter 1—Christian Harder, Clint Brown
Chapter 2—Mark Harrower, Clint Brown
Chapter 3—Allen Carroll, Rupert Essinger
Chapter 4—Christian Harder, Tamara Yoder
Chapter 5—Linda Beale, Andy Mitchell
Chapter 6—Nathan Shephard
Chapter 7—Will Crick, Justin Colville
Chapter 8—Christian Harder, Clint Brown
Chapter 9—Greg Tieman, Morakot Pilouk
Chapter 10—Clint Brown

The Learn ArcGIS team is Riley Peake, Bradley Wertman, Brandy Perkins, Colin Childs, John Berry, Kyle Bauer, and Veronica Rojas.

Special thanks to Catherine Ortiz, Eleanor Haire, Monica McGregor, Kylie Donia, Jeff Shaner, Tammy Johnson, Molly Zurn, Robert Garrity, Kathryn Keranen, Lyn Malone, Sanjib Panda, Brian Peterson, Maria Lomoro, and Cliff Crabbe for their support throughout the project; and to Deane Kensok, Sean Breyer, and Adam Mollenkopf.

Editing:	Dave Boyles
	Carolyn Schatz
	Matt Artz
Book design and layout:	Christian Harder
Website design:	Bradley Wertman
Product planning:	Sandi Newman
Print production:	Lilia Arias

The work of a number of Esri cartographers and data scientists is featured. Thanks to Kenneth Field, Andrew Skinner, Wesley Jones, Michael Dangermond, Jim Herries, Lee Bock, Cooper Thomas, Marjean Pobuda, Lauren Bennett, Flora Vale, Earl Nordstrand, Lauren Scott-Griffin, Jennifer Bell, Lisa Berry, Charlie Frye, Owen Evans, Richie Carmichael, Suzanne Foss, John Nelson, Daniel Siegel, Steve Heidelberg, Keith VanGraafeiland, Brian Sims, Craig McCabe, Julia Holtzclaw, and Esri-UK.

Finally, thanks to the worldwide GIS user community for doing amazing work with ArcGIS technology.

Credits

68	Space time traffic: Florida Department of Highway Safety and Motor Vehicles, Brevard County Property Appraiser, Florida Fish and Wildlife Conservation Commission, Fish and Wildlife Research Institute.
68	Crimes in San Francisco: Esri.
68	Public Transit: Esri Living Atlas Team, MARTA data.
69	Crimes in Chicago: Esri.
69	Southeast drought: NOAA.
70	Green infrastructure: Esri.
70	GeoPlanner: Esri.
70	GeoDescriber: Michael Dangermond, Esri.
71	A River Reborn: Map data: National Park Service, USGS National Hydrological Dataset, Natural Earth. Photos courtesy of Beda Calhoun, Matt Stoecker, and the folks at Damnation.
72	Solar potential 3D visualization: City of Naperville, Illinois, and Esri.
72	Crop Health: Esri Learn ArcGIS team.
72	Decisive Moments in the Battle of Gettysburg: Esri, HERE, DeLorme, increment p, Intermap, USGS, METI/NASA, EPA, USDA.
73	Infographic: Esri.
74	Infographic: Esri.
75	Chicago Poverty: Data and Maps for ArcGIS.
76	Charting in ArcGIS Pro: Esri.
77	Insights screen capture: Esri.
78	Insights screen capture: Esri.
78	Video: Esri.
79–80	ModelBuilder models: Esri.
80	Mountain lion: photo by Steve Engleberg (CC 3.0).
81	Infographic: Esri.
84	Mountain lion: photo courtesy of National Geographic.
85–86	3D Mountains: Esri, DigitalGlobe, Earthstar Geographics, CNES/Airbus DS, GeoEye, USDA FSA, USGS, Getmapping, Aerogrid, IGN, IGP, and the GIS User Community.
87	Peaks and Valleys: Esri Story Maps team.
88	Esri 3D Campus: Esri.
88	3D story map: Esri featuring CityEngine data.
88	3D Portland: Esri featuring CityEngine data.
89	Magellan's Route: Originally published for the Thinking Spatially Using GIS for desktop use by Esri Press.
89	3D Indianapolis: Esri 3D team.
90	Interesting Places: Esri, DigitalGlobe, Earthstar Geographics, CNES/Airbus DS, GeoEye, USDA FSA, USGS, Getmapping, Aerogrid, IGN, IGP, swisstopo, and the GIS User Community.
90	Recent earthquakes in 3D: Esri, USGS.
91	Scheidam: Nathan Shephard, Esri 3D team.
91	Marseilles: Nathan Shephard, Esri 3D team.
92	Calimesa: Christian Harder, Esri.
92	Quebec Tunnel: Control System.
93	Montreal: City of Montreal, Canada, Esri Canada.
93	3D bike ride: Brian Sims, Esri.
93	Oculus Rift video: Esri CityEngine team.
94	Philadelphia Visibility: Esri CityEngine team.
94	Typhoons: Nathan Shephard, Esri 3D team.
95	Street scene: Nathan Shephard, Esri 3D team.
95	Nathan Shephard: photo by Esri.
96	3D Portland Development: Oregon Metro, City of Portland, Esri.
96	3D Building routing: Esri 3D team.
96	Chicago Narcotics: Nathan Shephard, Esri 3D team.
97	Esri Developer Summit Video screen capture: Esri.
98	Miami shoreline Esri Learn team.
99–100	Arctic DEM Explorer: National Geospatial-Intelligence Agency (NGA), The National Science Foundation (NSF), and University of Minnesota's Polar Geospatial Center (PGC).
101	Peaks and Valleys: Esri Story Maps team.
102	Seville Art: Sevilla, Spain, sevilla.org.
102	CCTV Observations: Esri, City of Naperville, Illinois.
103	Downtown Reborn: City of Greenville, South Carolina.
104	iGeology: iGeology, British Geological Society.
104	LA Clean Streets: City of Los Angeles, Esri.
104	Opioid Crowdsource: Jeremiah Lindemann, Esri.
105	Pronatura Noroeste: Pronatura, Noroeste.
105	Kalmar Museum: Esri, Kalmar County, Sweden.
106	Drone2Map video screen capture: Chris Lesueur, Esri.
107	Operations dashboard screen capture: Esri.
107	Avalanche story map: ALErT: Anatolian pLateau climatE and Tectonic hazards, an EU-funded initiative.

108	Arctic DEM: NSF and The University of Minnesota's Polar Geospatial Center (PGC), Esri.
108	USDA Forest Service: USDA Forest Service, Esri.
108	Tapestry app: Esri, Data and Maps for ArcGIS.
109	Donegal Hills: County Donegal Map Portal.
109	GIS in Excel: Esri Maps for Office team.
109	GeoPlanner: Esri GeoPlanner team, Bill Miller, Esri.
110	Jeff Shanner: photo by Esri.
111	Explorer video screen capture: Esri.
111	Hen Harrier: Hen Harrier Special Protection Area (SPA) Habitat Mapping project.
111	FAA Data viewer: Federal Aviation Administration.
112	USGS Historical Topographic Map Explorer: USGS, Esri.
114	Fireman at hydrant: photo by eralt, (CC for 3.0).
115–16	Cubism Landsat Style: USGS Earth as Art 4 gallery 4.
117	Stereoscopic imagery: photo from Operation Crossbow: How 3D Glasses Helped Defeat Hitler: Source - BBC News / BBC Sport / bbc.co.uk - © 2011 BBC.
118	The Blue Marble: NASA/Apollo 17 crew; taken by either Harrison Schmitt or Ron Evans.
	Buzz Aldrin standing on the moon: NASA/Apollo 11 crew; taken by Neil Armstrong.
118	Neil Armstrong: One Small Step First Walk on Moon: by thenatman.
119	Landsat Shaded basemap: Esri, USGS, NASA.
119	Landsat: Unlocking Earth's Secrets: Esri, HERE, DeLorme, NGA, USGS, NASA.
120	High-resolution imagery: Esri, Earthstar Geographics, HERE, DeLorme.
121	Breathing Ranges: Esri, Visible Earth, NASA.
121	The Moisture Index: How Wet or Dry?: Esri, HERE, DeLorme, FAO, NOAA, USGS.
122	Severe Weather Public Information Map: Esri, HERE, DeLorme, FAO, NOAA, USGS, EPA, NPS	AccuWeather, Inc.	© 2013 Esri.
122	San Francisco 1859 and Today: Esri, SFEI & Quantum Spatial, USDA FSA, Microsoft, David Rumsey Historical Map Collection.
125	World Land Cover: Esri, HERE, DeLorme, NGA, USGS	Source: MDAUS	Esri, HERE.
126	Interesting Places: Esri, USDA, FSA, Microsoft.
126	A Hillshade Everyone Can Use: Esri, USGS, NOAA, DeLorme, NPS, CGIAR.
126	Pictometry 3D Scene: Esri, USDA, FSA, Microsoft.
127	Lena River Delta, Russia: USGS, Landsat NASA, Esri, Earthstar Geographics, CNES/Airbus DS.
128	Colima Volcano: Esri, Earthstar Geographics, CNES/Airbus DS, USGS, NASA.
128	Himalayas: Esri, Earthstar Geographics, CNES/Airbus DS.
128	Negro River, Brazil: Esri, Earthstar Geographics, CNES/Airbus DS.
129	Earthrise: NASA.
130	Discovering Liquid Water on Mars: Esri, NASA, JPL-Caltech, Univ. of Arizona.
130	New Horizons: Revealing Pluto's Secrets: Esri, NASA.
130	(Is there) Life on Mars?: Ken Field, Esri, Mars Orbiter Laser Altimeter (MOLA) instrument on the MGS (NASA/JPL/GSF) at approximately 463m/px. Official IAU/ USGS approved nomenclature from the MRCTR GIS lab (USGS). Landing site data from NASA.
132	Getting Started with Imagery, Learn ArcGIS: Esri.
133–34	IoT spread: Christian Harder, Esri.
135	Current wind and weather: METAR/TAF data provided from NOAA.
135	Misery Map: courtesy of FlightAware.
136	FedEx map (video capture): FedEx and Esri.
137	USA Flood Map: National Water Information System, Esri.
139	Suzanne Foss and Adam Mollenkopf: photo by Esri.
140	Operations Dashboard: Esri.
142	Blanco River flood: San Marcos, Texas.
143	Grid of 9 images: The Living Atlas of the World screen captures, Esri.
144	FIMAN: Flood Inundation and Mapping Network.
147–48	GIS Is about Community spread: Esri, photos by Esri.
149	Esri User Conference: photo by Esri.
150	Clint Brown: photo by Esri.
155	Jack Dangermond @ TED Conference: courtesy of TED.
156	Lauren Bennett: photo by Esri.
156	Beck's Hybrid Case Study: Beck's Hybrid, Esri.
157	Tree Survey crowdsource app: Esri.
158	Cumberland Colors: Municipality of Cumberland, Nova Scotia, Canada.
158	1Frame4Nature: International League of Conservation Photographers.
158	Downtown Eugene: City of Eugene, Oregon.
159	Science Matters story map: Esri Story Maps team.
162	World population estimate: Esri Living Atlas of the World.
165	Usu Volcano—Toya Caldera: UNESCO Global Geopark, RRIM.
169	Open Opportunity Data: Esri, the White House, US Department of Housing and Urban Development (HUD) HUD eGIS.

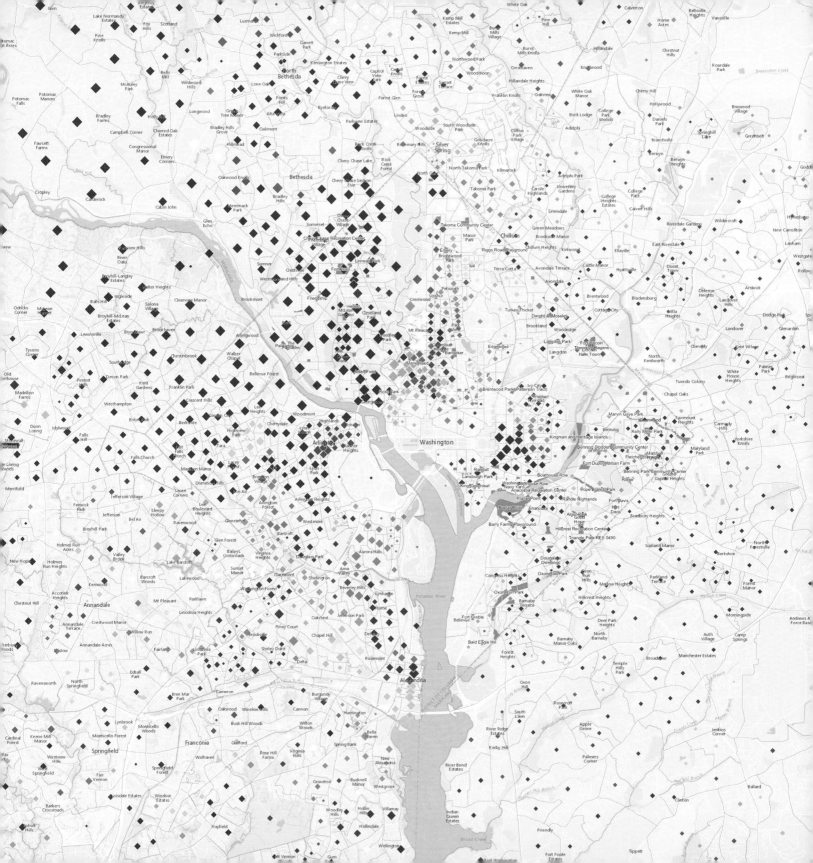